A BRIEF ATLAS OF THE SKELETON, SURFACE ANATOMY, AND SELECTED MEDICAL IMAGES

GERARD J. TORTORA

Bergen Community College

Cadaver Photographs *by* MARK NIELSEN

University of Utah

WILEY

John Wiley & Sons, Inc.

ISBN 978-0-470-14113-7

Printed in the United States of America

10 9 8 7 6 5 4 3 2 1

PREFACE

A Brief Atlas of the Skeleton, Surface Anatomy, and Selected Medical Images is offered to accompany any textbook of Anatomy and Physiology or Anatomy. It is unique among atlases currently available in terms of topic selection.

Part One • Bones of the Skeleton. This section provides a series of photographs of the axial and appendicular skeletons prepared by Mark Nielsen of the University of Utah. The photographs have been carefully selected, oriented, and labeled for maximum effectiveness as both a supplement to illustrations in textbooks and as a guide for learning the skeletal system in the laboratory. The photographs are enhanced by the inclusion of orientation diagrams that indicate how planes are used to make various sections and the direction from which a part of the body is viewed.

Part Two • Surface Anatomy. This part of the atlas provides a series of photos, by region, that illustrate the important landmarks on the exterior of the body. This information is important anatomically in locating structures such as bones, skeletal muscles, blood vessels, nerves, and internal organs. The information is important clinically for conducting physical examinations and performing certain diagnostic tests. Health care providers use their knowledge of surface anatomy to take pulse, measure blood pressure, draw blood, stop bleeding, insert needles and tubes, make surgical incisions, and listen to sounds made by the heart and lungs.

Part Three • Medical Images. A series of images provide students with an understanding of the many ways to study both normal anatomy and applied anatomy in clinical medicine for the diagnosis of disease. The images include CT and MRI scans, radiographs, micrographs, urograms, radionuclide scans, SPECT scans, PET scans, myelograms, angiograms, bronchograms, sonograms, mammograms, endoscopic views, and labaroscopic views.

ACKNOWLEDGEMENTS

There are many people whom I wish to thank for their efforts in helping me prepare this atlas. Bonnie Roesch, Executive Editor, my editor for so many of my publications, continues to provide the guidance, creativity, and professionalism which have impacted so profoundly on all of my books. There is no way that I can adequately express my gratitude to Bonnie. Clay Stone, Marketing Manager, has been a vital component of all my publications with Wiley. Clay coordinates a network of individuals who present the salient features of my books to professors. The feedback that Clay provides me from professors and students is invaluable. Kelly Tavares, Associate Production Manager, has once again distinguished herself as a "super" coordinator for all aspects of the production process. Her commitment to and passion for her job is so obvious and appreciated. Karin Gerdes Kincheloe, Senior Designer, is responsible for the cover and text design. Throughout, the pages are visually attractive, pedagogically effective, and student-friendly. For so many years, Claudia Durrell has been my art coordinator. Her organizational skills and artistic vision continue to amaze me. She always knows what to do and she does it so well. Hilary Newman, the photo manager, has pursued my photo requests with tenacity and has provided me with many options from which to choose. I also want to acknowledge Mary O'Sullivan, Project Editor, for her outstanding efforts in coordinating the development of supplements for my text, such as this atlas.

In preparing this atlas, I have strived to keep uppermost in my mind your needs as a student. As always, I could benefit from your comments and suggestions, which I hope you will send to me at the address below.

Gerard J. Tortora
Science and Technology, S229
Bergen Community College
400 Paramus Road
Paramus, NJ 07652

TABLE OF CONTENTS

Part Three • Medical Images

PART ONE:

BONES OF THE SKELETON

Divisions of the Adult Skeletal System			
Regions of the Skeleton	**Number of Bones**	**Regions of the Skeleton**	**Number of Bones**
Axial Skeleton		**Appendicular Skeleton**	
Skull		Pectoral (shoulder) girdles	
Cranium	8	*Clavicle*	2
Face	14	*Scapula*	2
Hyoid	1	Upper limbs (extremities)	
Auditory ossicles	6	*Humerus*	2
Vertebral column	26	*Ulna*	2
Thorax		*Radius*	2
Sternum	1	*Carpals*	16
Ribs	24	*Metacarpals*	10
	Subtotal = 80	*Phalanges*	28
		Pelvic (hip) girdle	
		Hip, pelvic, or coxal bone	2
		Lower limbs (extremities)	
		Femur	2
		Fibula	2
		Tibia	2
		Patella	2
		Tarsals	14
		Metatarsals	10
		Phalanges	28
			Subtotal = 126
			Total = 206

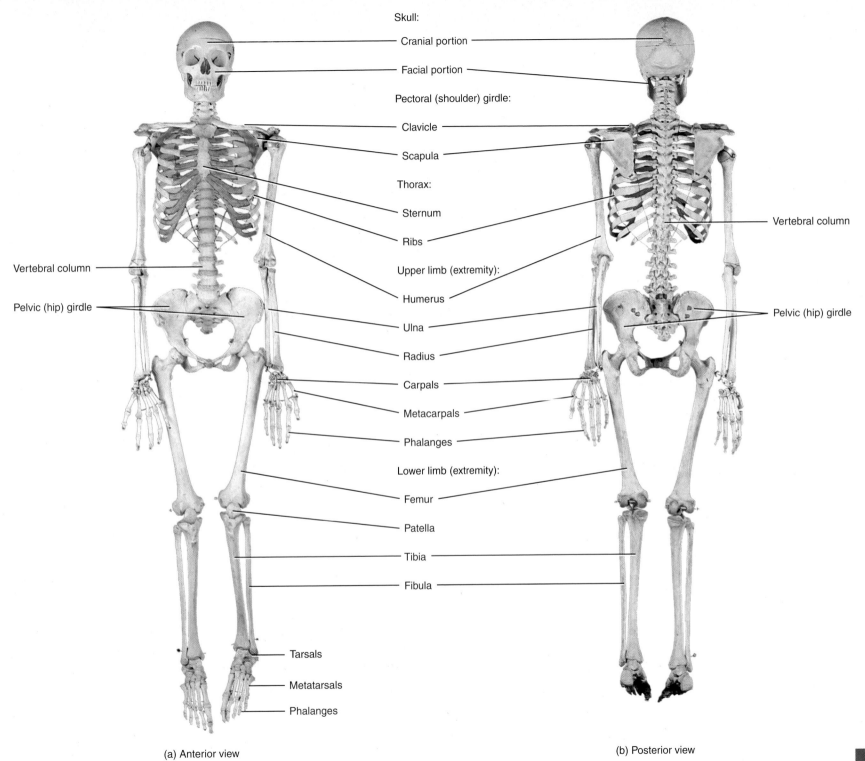

Skull:

Cranial portion

Facial portion

Pectoral (shoulder) girdle:

Clavicle

Scapula

Thorax:

Sternum

Ribs

Upper limb (extremity):

Humerus

Ulna

Radius

Carpals

Metacarpals

Phalanges

Lower limb (extremity):

Femur

Patella

Tibia

Fibula

Vertebral column

Pelvic (hip) girdle

Vertebral column

Pelvic (hip) girdle

Tarsals

Metatarsals

Phalanges

(a) Anterior view

(b) Posterior view

FIGURE 1.1 Complete skeleton.

1

SUPERIOR

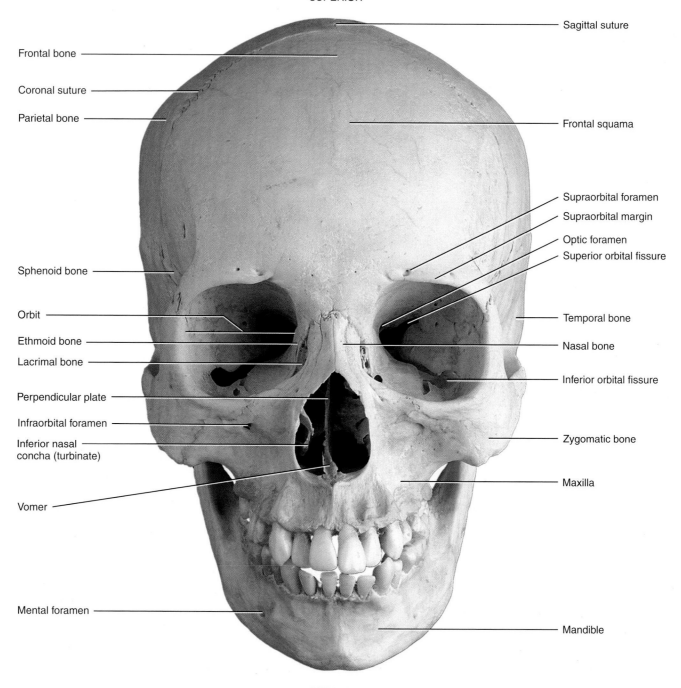

Sagittal suture

Frontal bone

Coronal suture

Parietal bone

Frontal squama

Supraorbital foramen

Supraorbital margin

Optic foramen

Superior orbital fissure

Sphenoid bone

Orbit

Temporal bone

Ethmoid bone

Nasal bone

Lacrimal bone

Inferior orbital fissure

Perpendicular plate

Infraorbital foramen

Zygomatic bone

Inferior nasal
concha (turbinate)

Maxilla

Vomer

Mental foramen

Mandible

INFERIOR

FIGURE 1.2 Skull, anterior view.

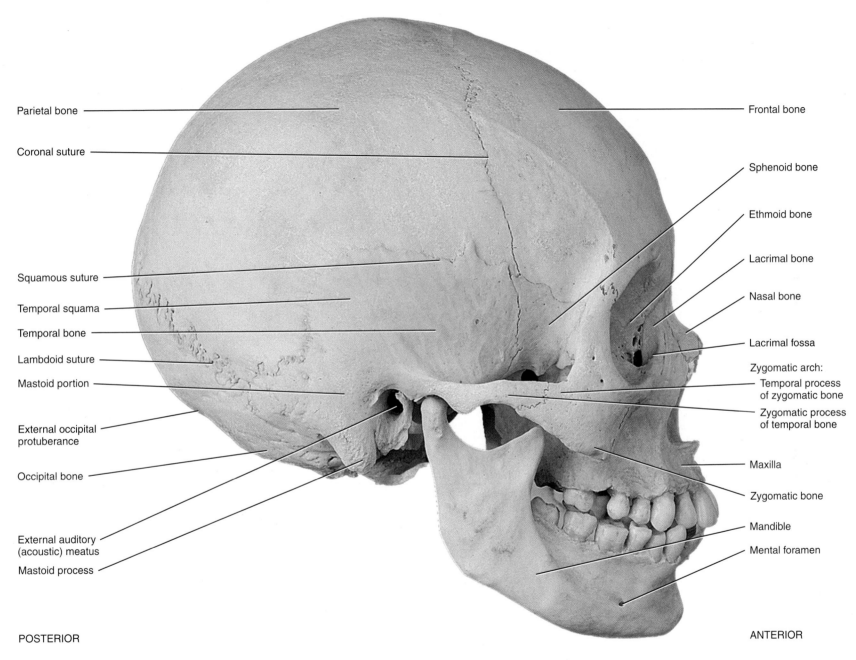

Parietal bone

Coronal suture

Squamous suture

Temporal squama

Temporal bone

Lambdoid suture

Mastoid portion

External occipital
protuberance

Occipital bone

External auditory
(acoustic) meatus

Mastoid process

Frontal bone

Sphenoid bone

Ethmoid bone

Lacrimal bone

Nasal bone

Lacrimal fossa

Zygomatic arch:

Temporal process
of zygomatic bone

Zygomatic process
of temporal bone

Maxilla

Zygomatic bone

Mandible

Mental foramen

POSTERIOR

ANTERIOR

FIGURE 1.3 Skull, right lateral view.

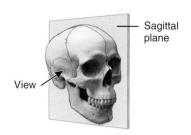

Sagittal plane

View

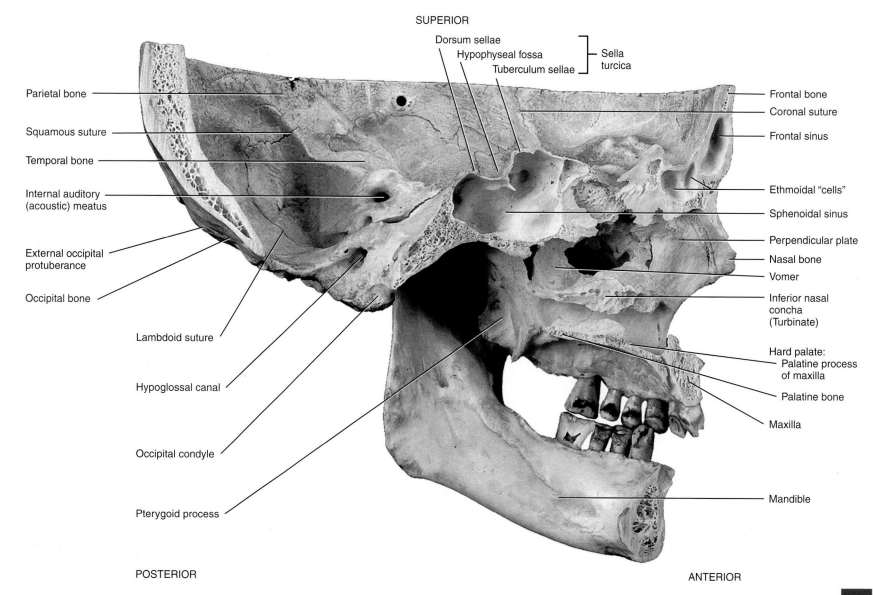

SUPERIOR

Dorsum sellae

Hypophyseal fossa

Tuberculum sellae

Sella turcica

Parietal bone

Squamous suture

Temporal bone

Internal auditory (acoustic) meatus

External occipital protuberance

Occipital bone

Lambdoid suture

Hypoglossal canal

Occipital condyle

Pterygoid process

Frontal bone

Coronal suture

Frontal sinus

Ethmoidal "cells"

Sphenoidal sinus

Perpendicular plate

Nasal bone

Vomer

Inferior nasal concha (Turbinate)

Hard palate:
Palatine process of maxilla

Palatine bone

Maxilla

Mandible

POSTERIOR

ANTERIOR

FIGURE 1.4 Skull, medial view of sagittal section.

4

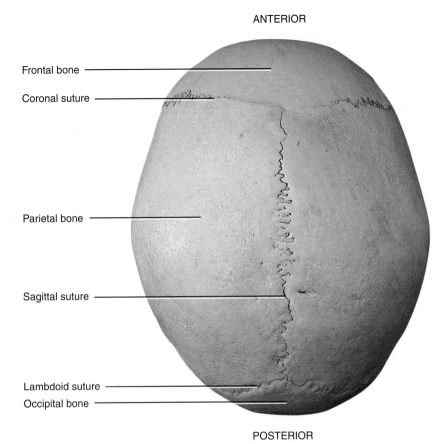

ANTERIOR

Frontal bone

Coronal suture

Parietal bone

Sagittal suture

Lambdoid suture

Occipital bone

POSTERIOR

FIGURE 1.6 Skull, superior view.

SUPERIOR

Sagittal suture

Parietal bone

Lambdoid suture

Occipital bone

External occipital protuberance

INFERIOR

FIGURE 1.5 Skull, posterior view.

View

ANTERIOR

Incisor teeth

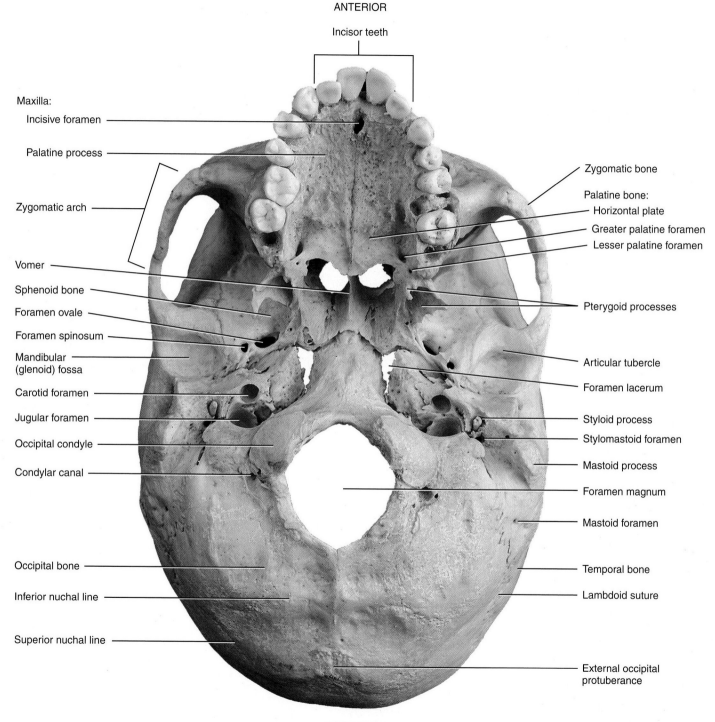

Maxilla:

Incisive foramen

Palatine process

Zygomatic arch

Vomer

Sphenoid bone

Foramen ovale

Foramen spinosum

Mandibular
(glenoid) fossa

Carotid foramen

Jugular foramen

Occipital condyle

Condylar canal

Occipital bone

Inferior nuchal line

Superior nuchal line

Zygomatic bone

Palatine bone:

Horizontal plate

Greater palatine foramen

Lesser palatine foramen

Pterygoid processes

Articular tubercle

Foramen lacerum

Styloid process

Stylomastoid foramen

Mastoid process

Foramen magnum

Mastoid foramen

Temporal bone

Lambdoid suture

External occipital
protuberance

POSTERIOR

FIGURE 1.7 Skull, inferior view.

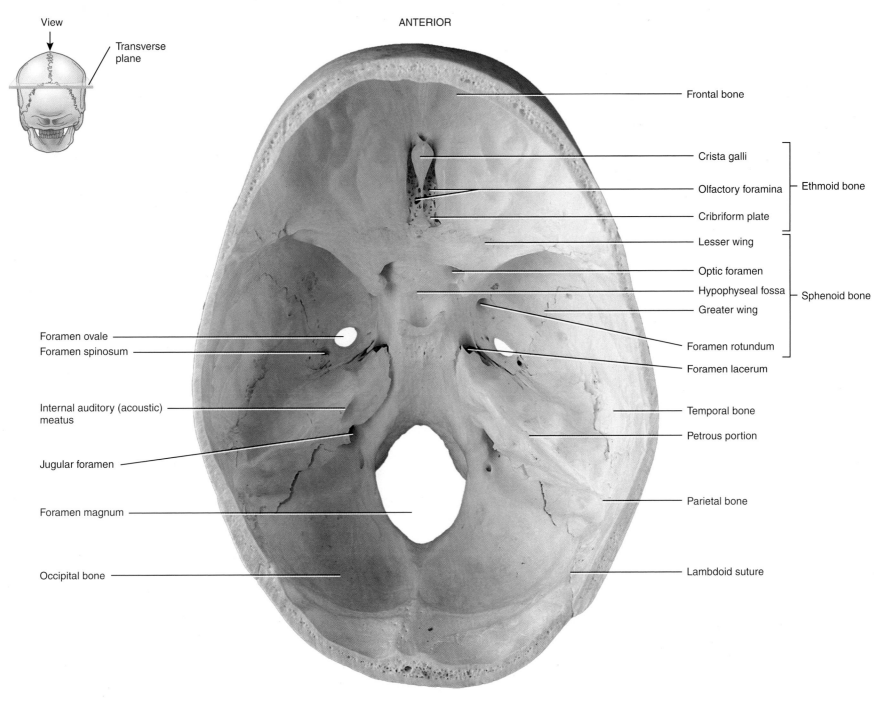

View

Transverse plane

ANTERIOR

Frontal bone

Crista galli

Olfactory foramina

Ethmoid bone

Cribriform plate

Lesser wing

Optic foramen

Hypophyseal fossa

Sphenoid bone

Greater wing

Foramen ovale

Foramen spinosum

Foramen rotundum

Foramen lacerum

Internal auditory (acoustic) meatus

Temporal bone

Petrous portion

Jugular foramen

Parietal bone

Foramen magnum

Occipital bone

Lambdoid suture

POSTERIOR

FIGURE 1.8 Floor of cranium, superior view.

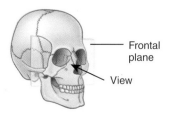

Frontal plane

View

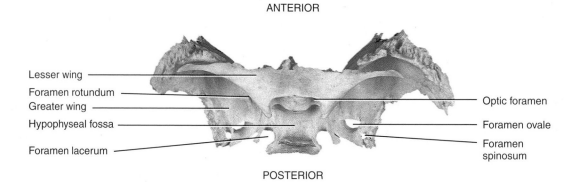

ANTERIOR

Lesser wing

Foramen rotundum

Greater wing

Hypophyseal fossa

Foramen lacerum

Optic foramen

Foramen ovale

Foramen spinosum

POSTERIOR

FIGURE 1.10 Superior view

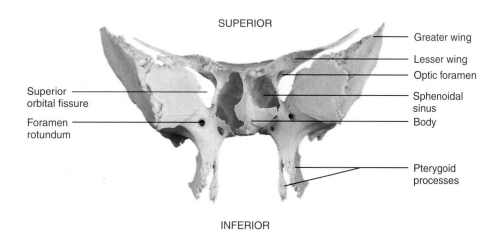

SUPERIOR

Greater wing

Lesser wing

Optic foramen

Sphenoidal sinus

Body

Superior orbital fissure

Foramen rotundum

Pterygoid processes

INFERIOR

FIGURE 1.9 Sphenoid bone, anterior view.

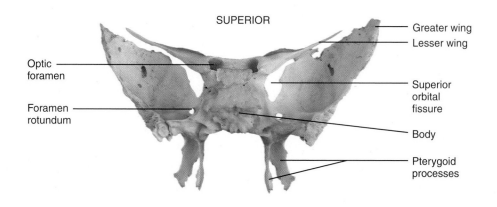

SUPERIOR

Greater wing

Lesser wing

Optic foramen

Foramen rotundum

Superior orbital fissure

Body

Pterygoid processes

INFERIOR

FIGURE 1.11 Sphenoid bone, posterior view.

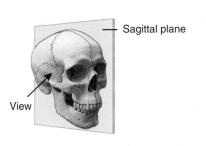

Sagittal plane

View

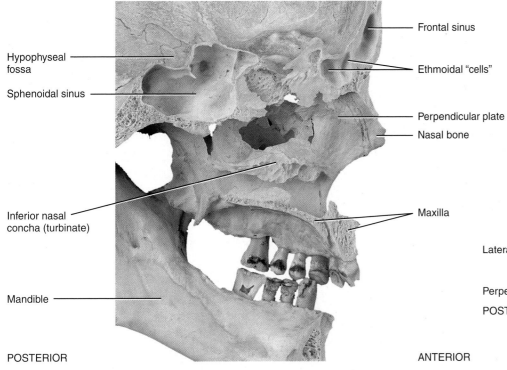

SUPERIOR

Frontal sinus

Hypophyseal fossa

Ethmoidal "cells"

Sphenoidal sinus

Perpendicular plate

Nasal bone

Inferior nasal concha (turbinate)

Maxilla

Mandible

POSTERIOR

ANTERIOR

FIGURE 1.12 Ethmoid bone, medial view of sagittal section.

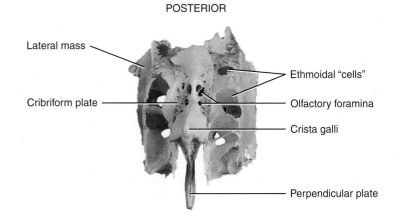

POSTERIOR

Lateral mass

Ethmoidal "cells"

Cribriform plate

Olfactory foramina

Crista galli

Perpendicular plate

ANTERIOR

FIGURE 1.13 Ethmoid bone, superior view.

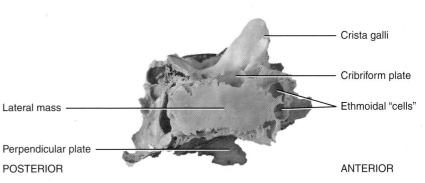

Crista galli

Cribriform plate

Ethmoidal "cells"

Lateral mass

Perpendicular plate

POSTERIOR

ANTERIOR

FIGURE 1.14 Ethmoid bone, right lateral view.

SUPERIOR

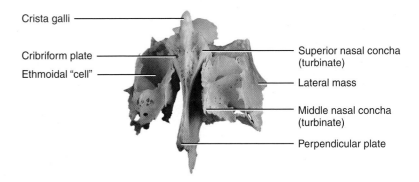

Crista galli

Cribriform plate

Ethmoidal "cell"

Superior nasal concha (turbinate)

Lateral mass

Middle nasal concha (turbinate)

Perpendicular plate

INFERIOR

FIGURE 1.15 Ethmoid bone, anterior view.

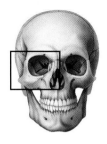

SUPERIOR

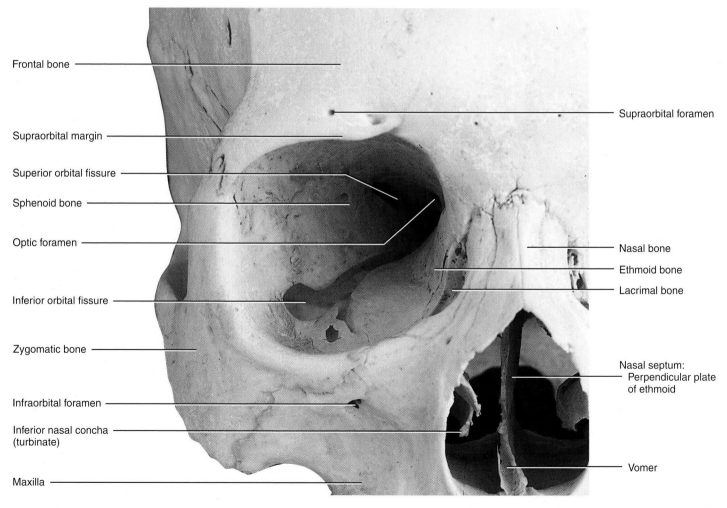

Frontal bone

Supraorbital foramen

Supraorbital margin

Superior orbital fissure

Sphenoid bone

Optic foramen

Nasal bone

Ethmoid bone

Lacrimal bone

Inferior orbital fissure

Zygomatic bone

Nasal septum:
    Perpendicular plate
    of ethmoid

Infraorbital foramen

Inferior nasal concha
(turbinate)

Vomer

Maxilla

INFERIOR

FIGURE 1.16 Right orbit, anterior view.

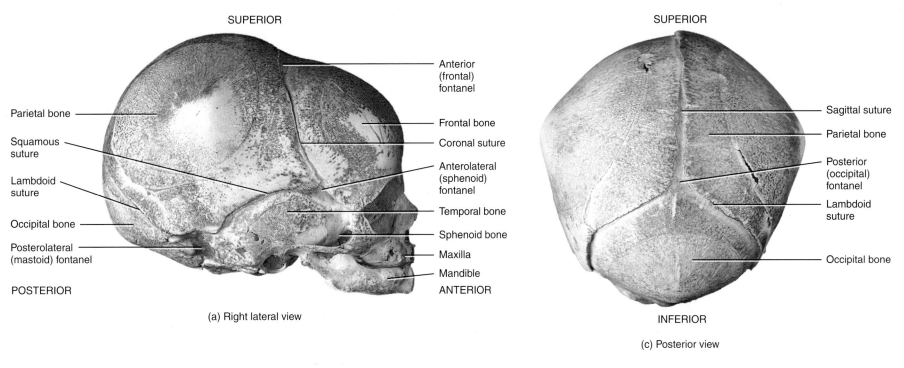

SUPERIOR

Parietal bone

Squamous
suture

Lambdoid
suture

Occipital bone

Posterolateral
(mastoid) fontanel

POSTERIOR

Anterior
(frontal)
fontanel

Frontal bone

Coronal suture

Anterolateral
(sphenoid)
fontanel

Temporal bone

Sphenoid bone

Maxilla

Mandible

ANTERIOR

(a) Right lateral view

SUPERIOR

Sagittal suture

Parietal bone

Posterior
(occipital)
fontanel

Lambdoid
suture

Occipital bone

INFERIOR

(c) Posterior view

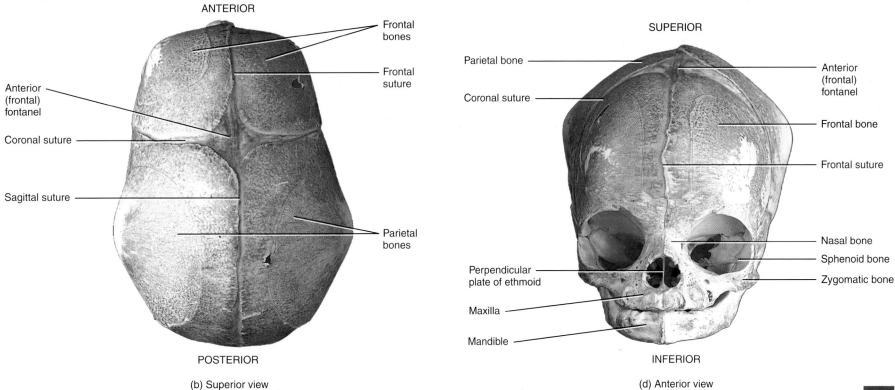

ANTERIOR

Frontal
bones

Frontal
suture

Anterior
(frontal)
fontanel

Coronal suture

Sagittal suture

Parietal
bones

POSTERIOR

(b) Superior view

SUPERIOR

Parietal bone

Coronal suture

Anterior
(frontal)
fontanel

Frontal bone

Frontal suture

Nasal bone

Sphenoid bone

Zygomatic bone

Perpendicular
plate of ethmoid

Maxilla

Mandible

INFERIOR

(d) Anterior view

FIGURE 1.17 Fetal skull, fontanels.

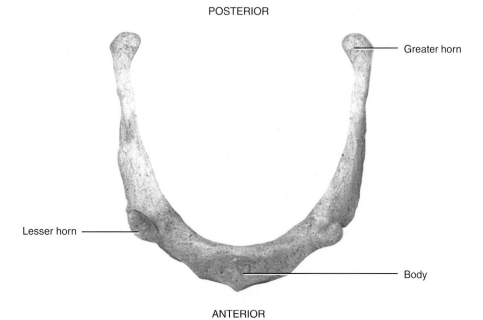

POSTERIOR

Greater horn

Lesser horn

Body

ANTERIOR

FIGURE 1.18 Hyoid bone

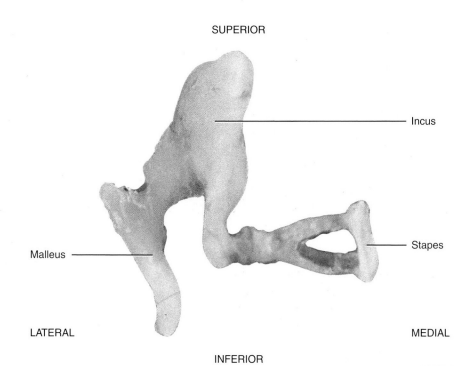

SUPERIOR

Incus

Malleus

Stapes

LATERAL

MEDIAL

INFERIOR

FIGURE 1.19 Auditory ossicles

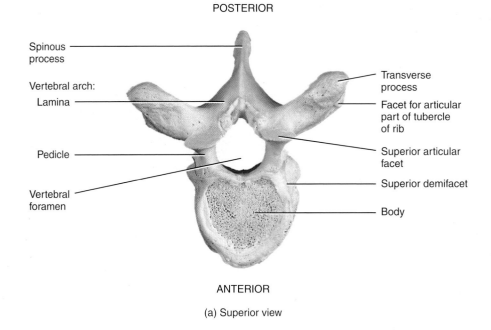

(a) Superior view

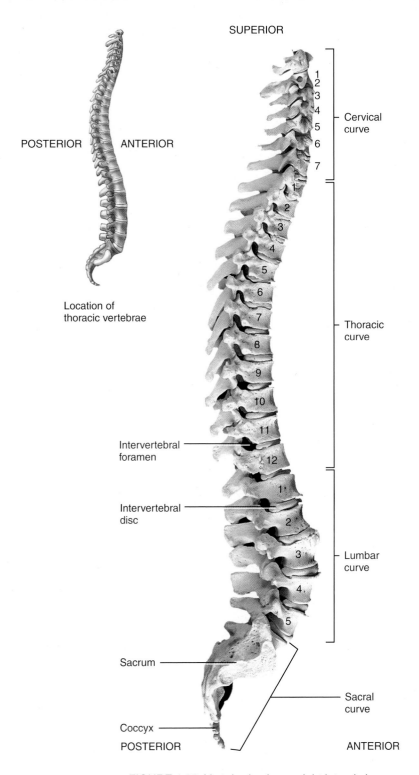

FIGURE 1.20 Vertebral column, right lateral view.

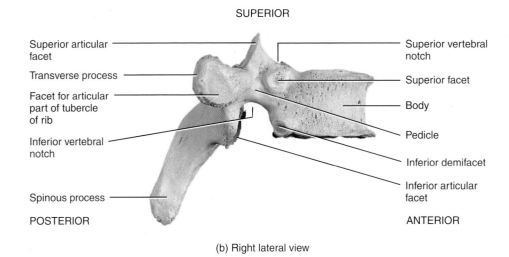

(b) Right lateral view

FIGURE 1.21 A typical (thoracic) vertebra.

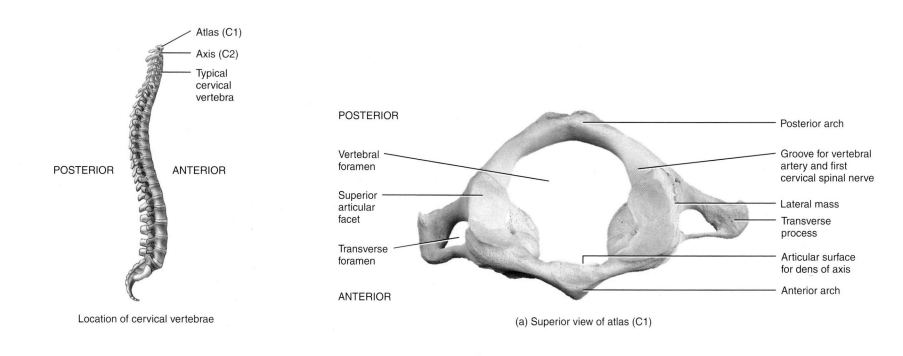

Atlas (C1)

Axis (C2)

Typical cervical vertebra

POSTERIOR   ANTERIOR

Location of cervical vertebrae

POSTERIOR

Vertebral foramen

Superior articular facet

Transverse foramen

ANTERIOR

Posterior arch

Groove for vertebral artery and first cervical spinal nerve

Lateral mass

Transverse process

Articular surface for dens of axis

Anterior arch

(a) Superior view of atlas (C1)

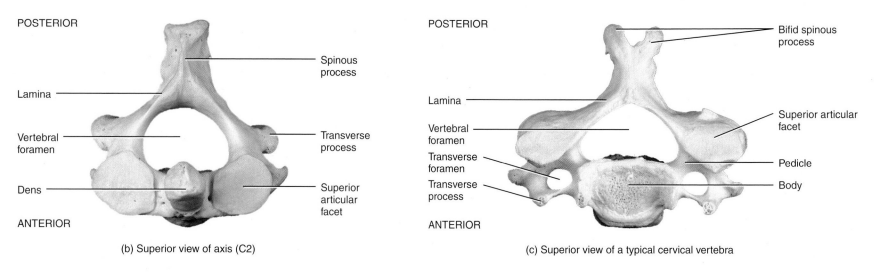

POSTERIOR

Lamina

Vertebral foramen

Dens

ANTERIOR

Spinous process

Transverse process

Superior articular facet

(b) Superior view of axis (C2)

POSTERIOR

Lamina

Vertebral foramen

Transverse foramen

Transverse process

ANTERIOR

Bifid spinous process

Superior articular facet

Pedicle

Body

(c) Superior view of a typical cervical vertebra

FIGURE 1.22 Cervical vertebrae.

14

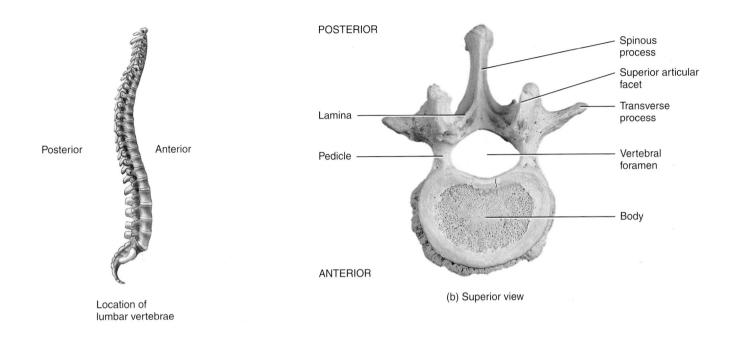

Posterior   Anterior

Location of
lumbar vertebrae

POSTERIOR

Spinous
process

Superior articular
facet

Lamina

Transverse
process

Pedicle

Vertebral
foramen

Body

ANTERIOR

(b) Superior view

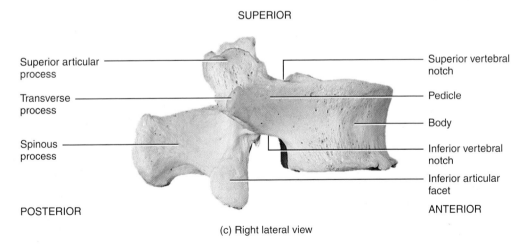

SUPERIOR

Superior articular
process

Superior vertebral
notch

Transverse
process

Pedicle

Body

Spinous
process

Inferior vertebral
notch

Inferior articular
facet

POSTERIOR   ANTERIOR

(c) Right lateral view

FIGURE 1.23 Lumbar vertebrae.

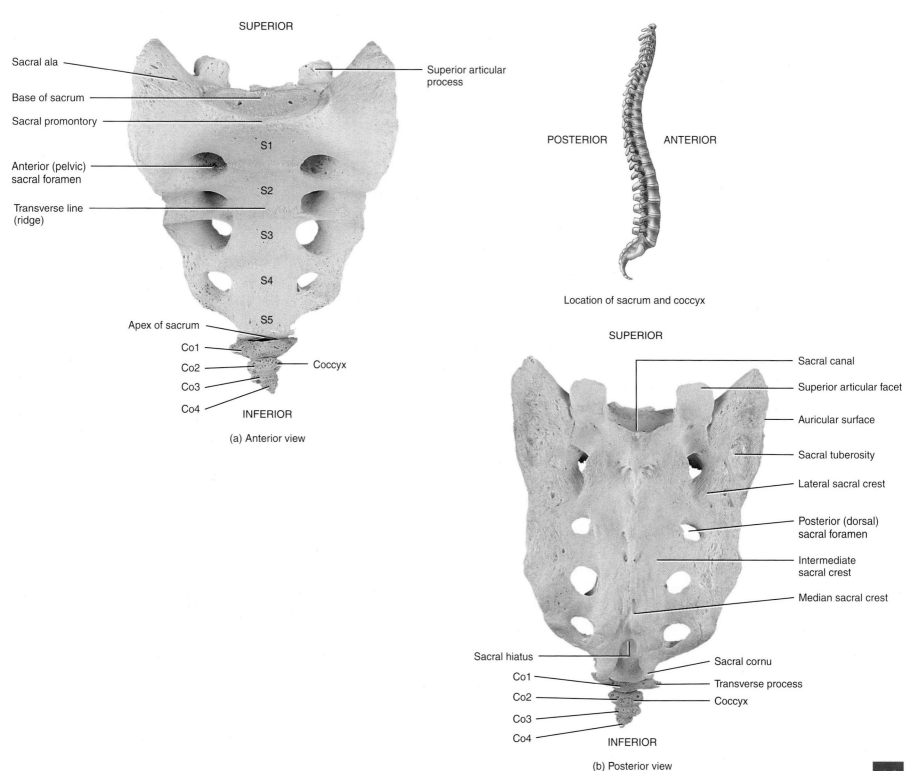

SUPERIOR

Sacral ala

Base of sacrum

Sacral promontory

S1

Anterior (pelvic) sacral foramen

S2

Transverse line (ridge)

S3

S4

S5

Apex of sacrum

Co1

Co2

Co3

Co4

Superior articular process

INFERIOR

Coccyx

(a) Anterior view

POSTERIOR   ANTERIOR

Location of sacrum and coccyx

SUPERIOR

Sacral canal

Superior articular facet

Auricular surface

Sacral tuberosity

Lateral sacral crest

Posterior (dorsal) sacral foramen

Intermediate sacral crest

Median sacral crest

Sacral hiatus

Co1

Co2

Co3

Co4

Sacral cornu

Transverse process

Coccyx

INFERIOR

(b) Posterior view

FIGURE 1.24 Sacrum and coccyx.

16

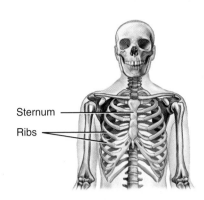

Sternum

Ribs

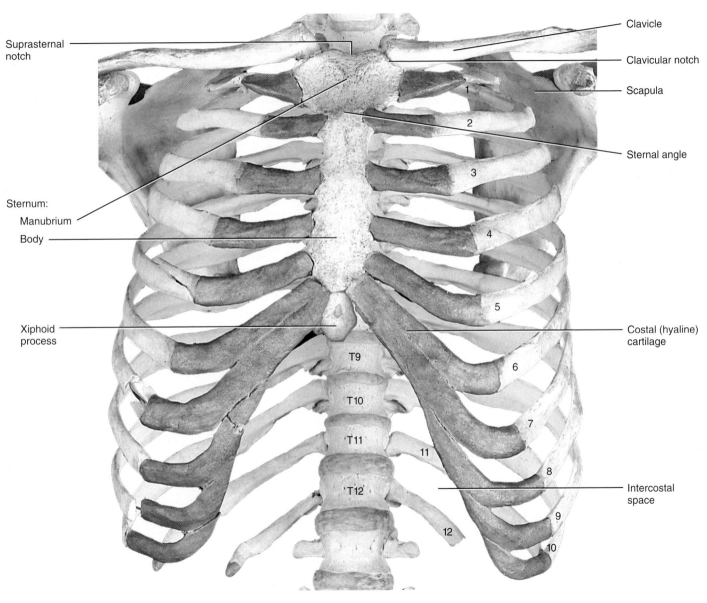

SUPERIOR

Suprasternal
notch

Sternum:
Manubrium

Body

Xiphoid
process

Clavicle

Clavicular notch

Scapula

Sternal angle

1

2

3

4

5

6

7

8

9

10

11

12

T9

T10

T11

T12

Costal (hyaline)
cartilage

Intercostal
space

INFERIOR

FIGURE 1.25 Skeleton of thorax, anterior view.

17

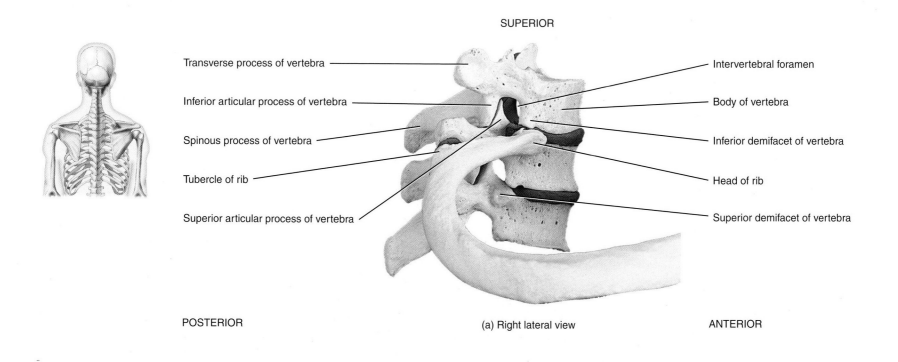

SUPERIOR

Transverse process of vertebra

Inferior articular process of vertebra

Spinous process of vertebra

Tubercle of rib

Superior articular process of vertebra

Intervertebral foramen

Body of vertebra

Inferior demifacet of vertebra

Head of rib

Superior demifacet of vertebra

POSTERIOR

(a) Right lateral view

ANTERIOR

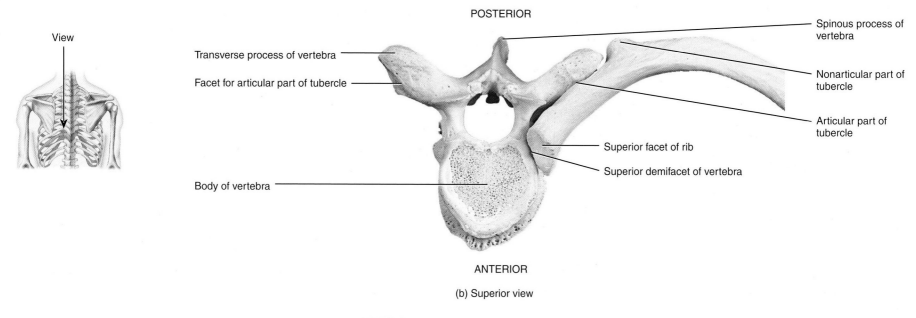

POSTERIOR

View

Transverse process of vertebra

Facet for articular part of tubercle

Body of vertebra

Spinous process of vertebra

Nonarticular part of tubercle

Articular part of tubercle

Superior facet of rib

Superior demifacet of vertebra

ANTERIOR

(b) Superior view

FIGURE 1.26 Articulation of thoracic vertebra with rib.

18

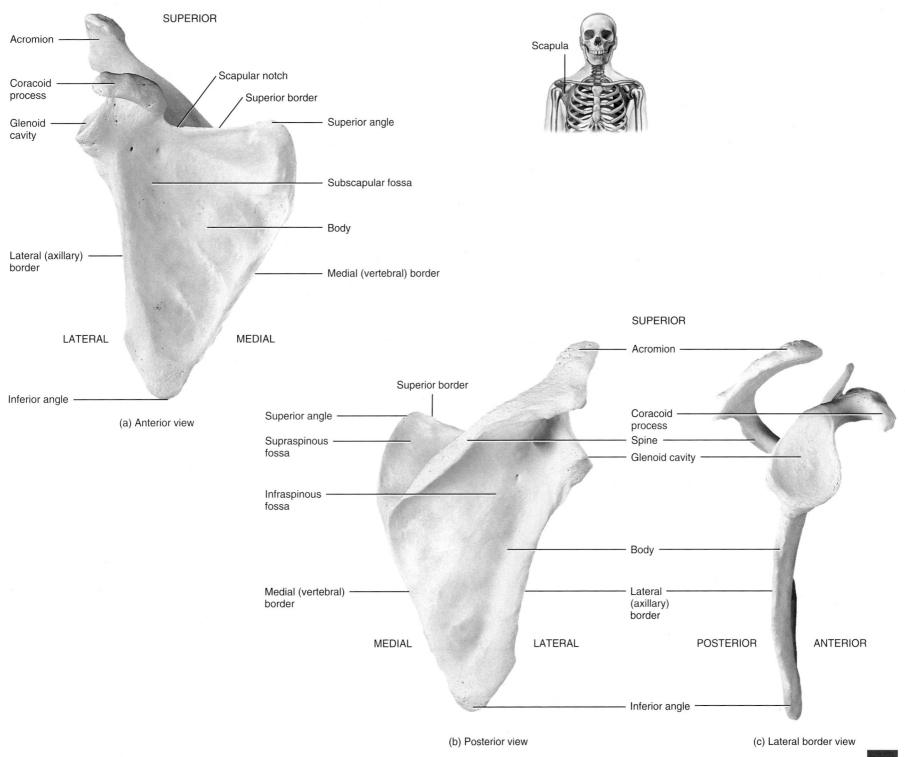

Acromion

Coracoid process

Glenoid cavity

SUPERIOR

Scapular notch

Superior border

Superior angle

Subscapular fossa

Body

Lateral (axillary) border

Medial (vertebral) border

LATERAL

MEDIAL

Inferior angle

(a) Anterior view

Scapula

Superior border

Superior angle

Supraspinous fossa

Infraspinous fossa

Medial (vertebral) border

MEDIAL

LATERAL

Inferior angle

(b) Posterior view

SUPERIOR

Acromion

Coracoid process

Spine

Glenoid cavity

Body

Lateral (axillary) border

POSTERIOR

ANTERIOR

(c) Lateral border view

FIGURE 1.27 Right scapula.

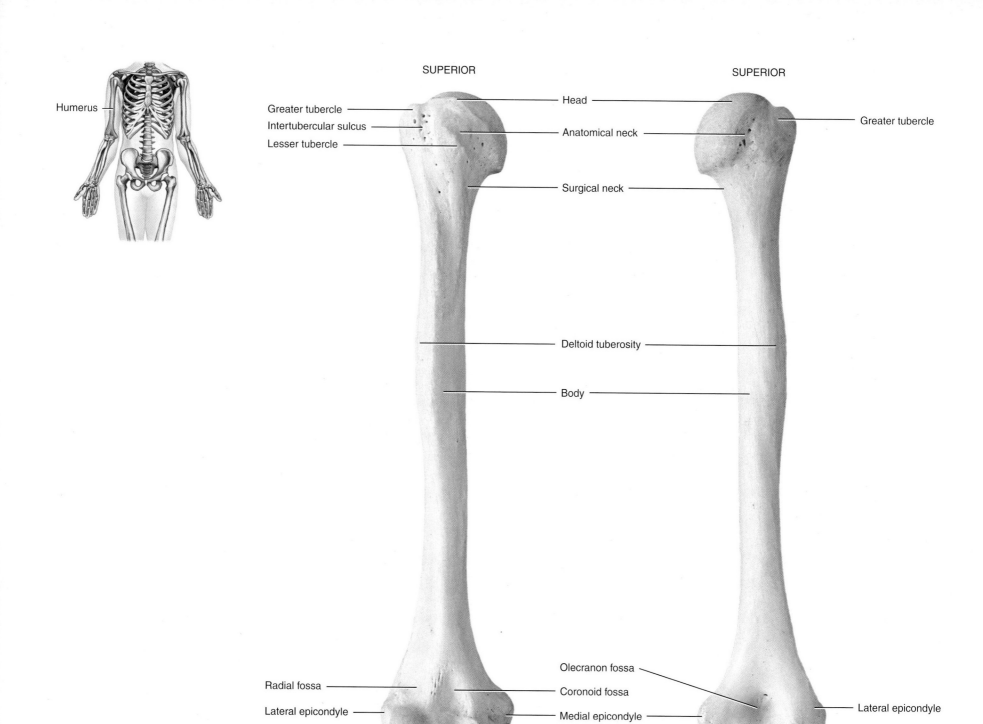

Humerus

SUPERIOR

Greater tubercle

Intertubercular sulcus

Lesser tubercle

Head

Anatomical neck

Surgical neck

Deltoid tuberosity

Body

SUPERIOR

Greater tubercle

Radial fossa

Lateral epicondyle

Capitulum

LATERAL

Olecranon fossa

Coronoid fossa

Medial epicondyle

Trochlea

MEDIAL

Lateral epicondyle

MEDIAL

LATERAL

(a) Anterior view

(b) Posterior view

FIGURE 1.28 Right humerus.

20

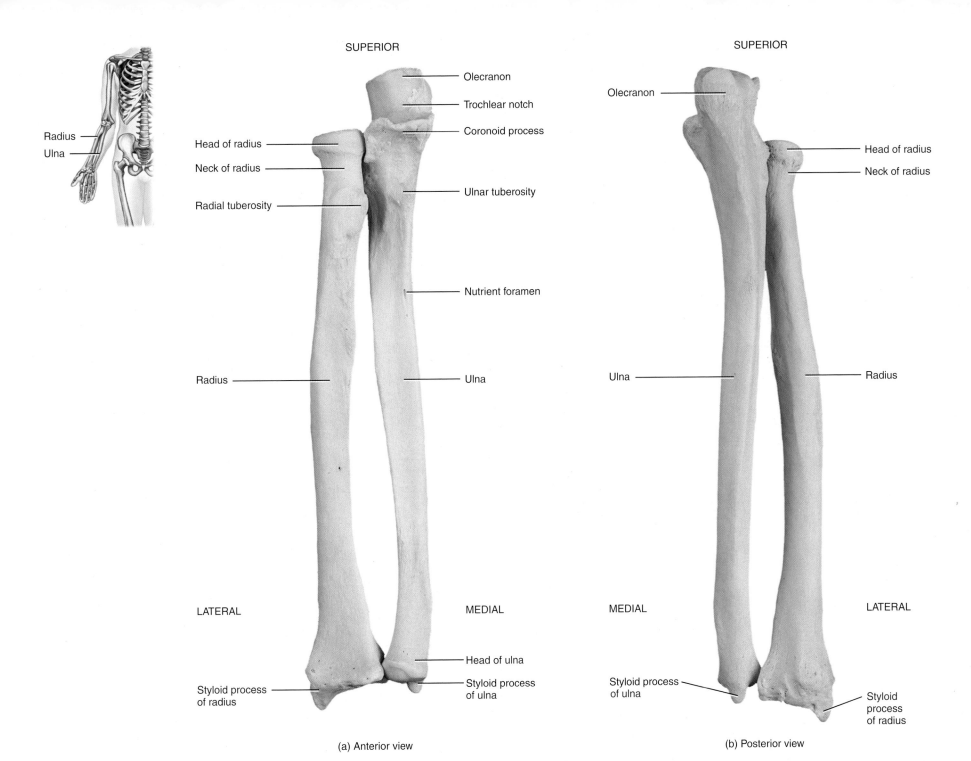

SUPERIOR

Radius
Ulna

SUPERIOR

Olecranon

Trochlear notch

Head of radius

Coronoid process

Neck of radius

Olecranon

Head of radius

Ulnar tuberosity

Neck of radius

Radial tuberosity

Nutrient foramen

Radius

Ulna

Ulna

Radius

LATERAL

MEDIAL

MEDIAL

LATERAL

Head of ulna

Styloid process
of radius

Styloid process
of ulna

Styloid process
of ulna

Styloid
process
of radius

(a) Anterior view

(b) Posterior view

FIGURE 1.29 Right ulna and radius.

21

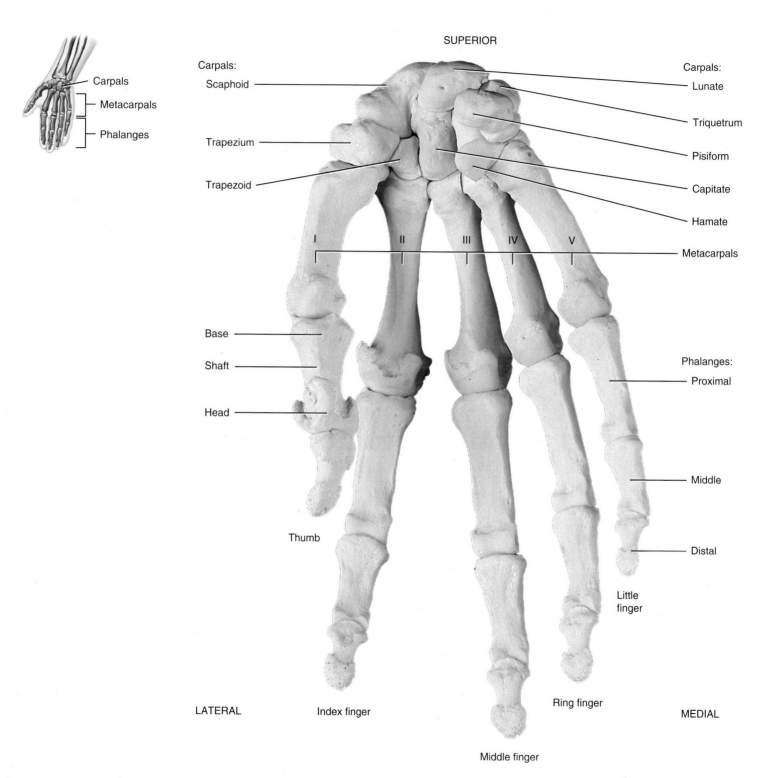

SUPERIOR

Carpals:

Carpals:

Scaphoid

Lunate

Triquetrum

Trapezium

Pisiform

Trapezoid

Capitate

Hamate

I   II   III   IV   V

Metacarpals

Base

Phalanges:

Shaft

Proximal

Head

Middle

Thumb

Distal

Little
finger

LATERAL

Index finger

Ring finger

MEDIAL

Middle finger

FIGURE 1.30 Right hand and wrist, anterior view.

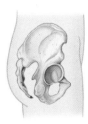

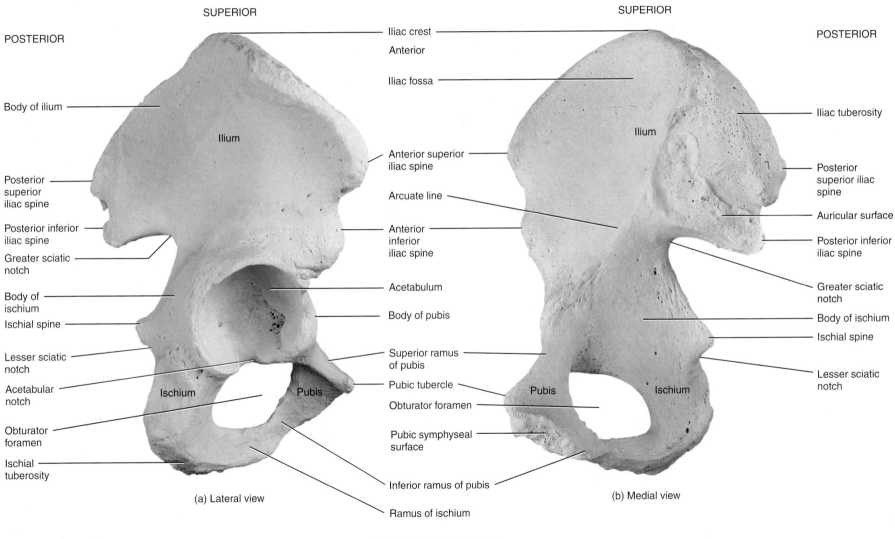

SUPERIOR

SUPERIOR

POSTERIOR

POSTERIOR

Iliac crest

Anterior

Iliac fossa

Body of ilium

Ilium

Iliac tuberosity

Ilium

Posterior superior iliac spine

Anterior superior iliac spine

Posterior superior iliac spine

Posterior inferior iliac spine

Arcuate line

Auricular surface

Greater sciatic notch

Anterior inferior iliac spine

Posterior inferior iliac spine

Body of ischium

Acetabulum

Greater sciatic notch

Ischial spine

Body of pubis

Body of ischium

Lesser sciatic notch

Superior ramus of pubis

Ischial spine

Acetabular notch

Ischium

Pubis

Pubic tubercle

Pubis

Ischium

Lesser sciatic notch

Obturator foramen

Obturator foramen

Pubic symphyseal surface

Ischial tuberosity

(a) Lateral view

(b) Medial view

Inferior ramus of pubis

Ramus of ischium

FIGURE 1.31 Right hip bone.

Pelvic (hip) girdle

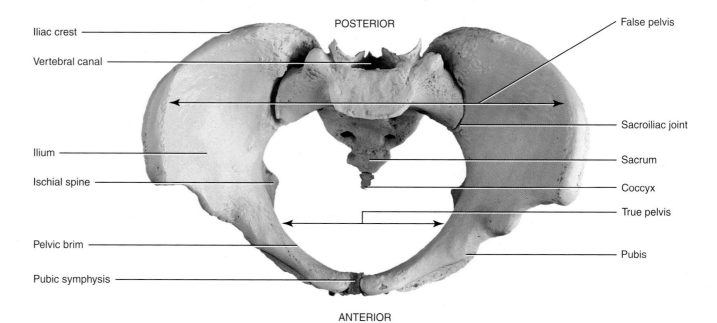

POSTERIOR

Iliac crest

Vertebral canal

Ilium

Ischial spine

Pelvic brim

Pubic symphysis

False pelvis

Sacroiliac joint

Sacrum

Coccyx

True pelvis

Pubis

ANTERIOR

(a) Female pelvis, superior view

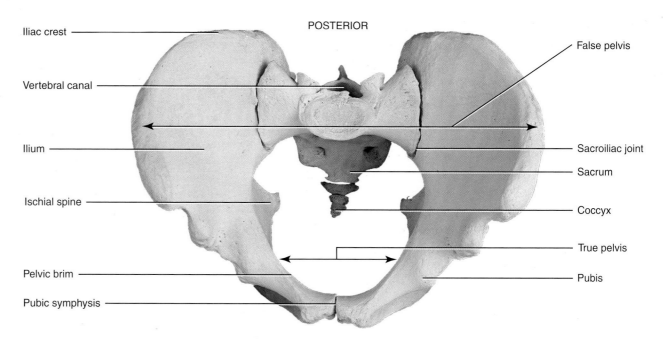

POSTERIOR

Iliac crest

Vertebral canal

Ilium

Ischial spine

Pelvic brim

Pubic symphysis

False pelvis

Sacroiliac joint

Sacrum

Coccyx

True pelvis

Pubis

ANTERIOR

(b) Male pelvis, superior view

FIGURE 1.32 Pelvis.

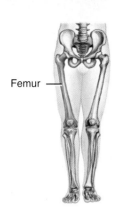

Femur

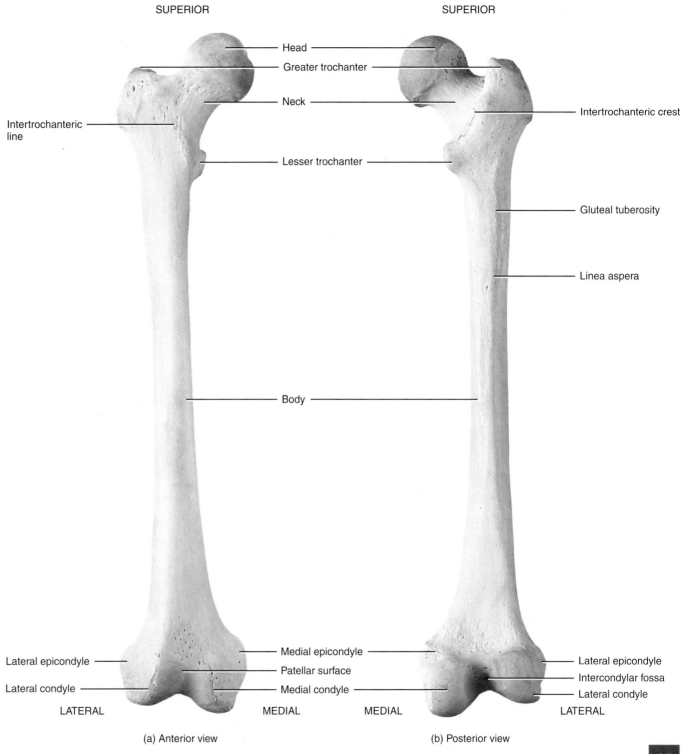

SUPERIOR

SUPERIOR

Head

Greater trochanter

Neck

Intertrochanteric crest

Intertrochanteric line

Lesser trochanter

Gluteal tuberosity

Linea aspera

Body

Medial epicondyle

Lateral epicondyle

Intercondylar fossa

Lateral epicondyle

Patellar surface

Lateral condyle

Medial condyle

Lateral condyle

LATERAL

MEDIAL

MEDIAL

LATERAL

(a) Anterior view

(b) Posterior view

FIGURE 1.33 Right femur.

25

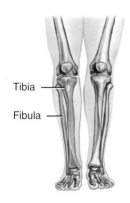

Tibia

Fibula

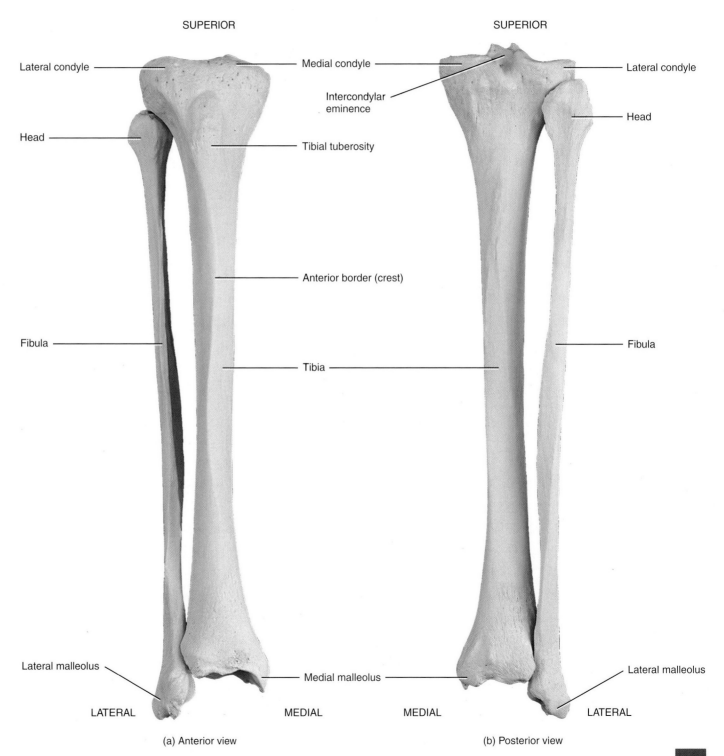

SUPERIOR

Lateral condyle

Head

Fibula

Lateral malleolus

LATERAL

Medial condyle

Intercondylar eminence

Tibial tuberosity

Anterior border (crest)

Tibia

Medial malleolus

MEDIAL

(a) Anterior view

SUPERIOR

Lateral condyle

Head

Fibula

Lateral malleolus

LATERAL

MEDIAL

(b) Posterior view

FIGURE 1.34 Right tibia and fibula.

Superior view

Tarsals
Metatarsals
Phalanges

LATERAL POSTERIOR MEDIAL

Tarsals:
Calcaneus

Talus

Tarsals:
Navicular

Cuboid

Cuneiforms:
Third (lateral)
Second (intermediate)
First (medial)

V IV III II I

Metatarsals

Base

Shaft

Head

Phalanges:
Proximal

Middle

Distal

Great (big) toe

ANTERIOR

FIGURE 1.35 Right ankle and foot, superior view.

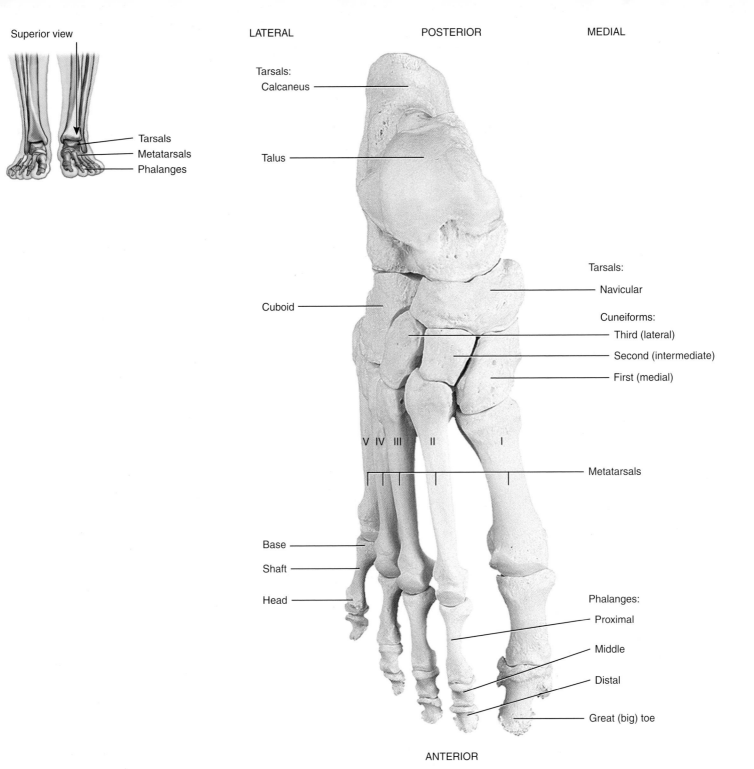

PART TWO:

SURFACE ANATOMY

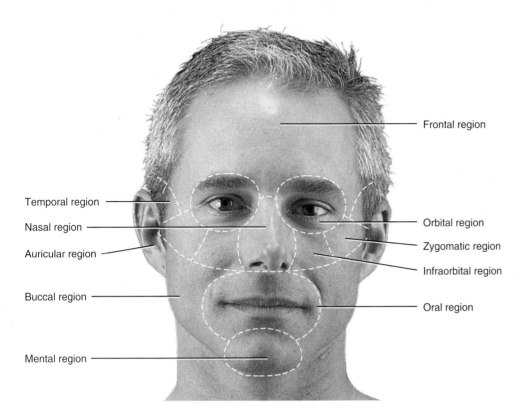

Frontal region

Temporal region

Nasal region

Auricular region

Buccal region

Mental region

Orbital region

Zygomatic region

Infraorbital region

Oral region

(a) Anterior view

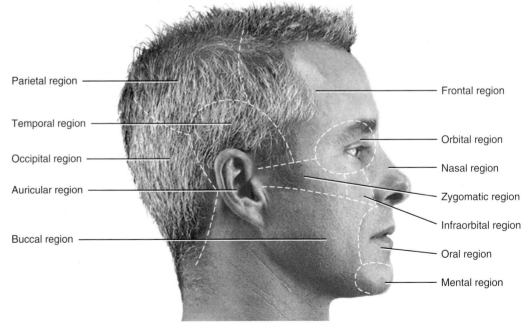

Parietal region

Temporal region

Occipital region

Auricular region

Buccal region

Frontal region

Orbital region

Nasal region

Zygomatic region

Infraorbital region

Oral region

Mental region

(b) Right lateral view

31

FIGURE 2.1 Principal regions of the cranium and face.

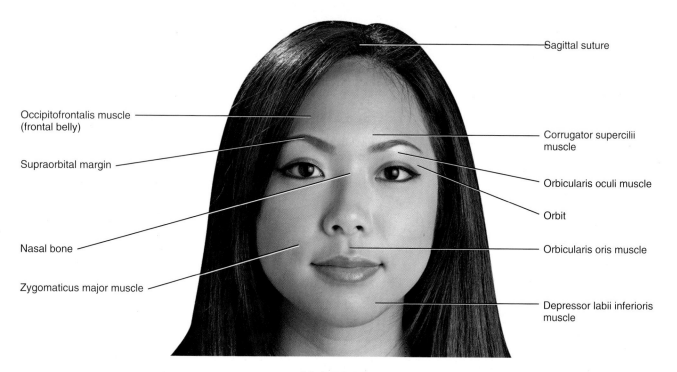

Occipitofrontalis muscle
(frontal belly)

Supraorbital margin

Nasal bone

Zygomaticus major muscle

Sagittal suture

Corrugator supercilii
muscle

Orbicularis oculi muscle

Orbit

Orbicularis oris muscle

Depressor labii inferioris
muscle

(a) Anterior view

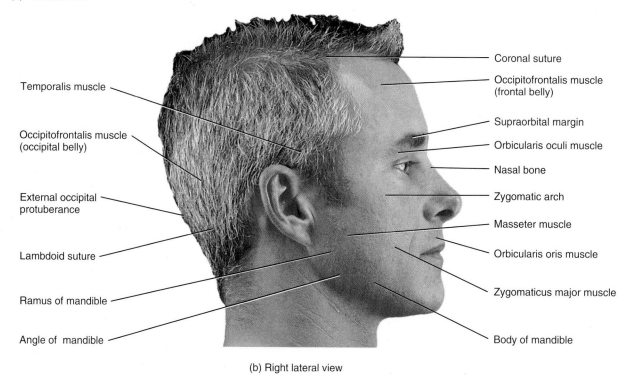

Temporalis muscle

Occipitofrontalis muscle
(occipital belly)

External occipital
protuberance

Lambdoid suture

Ramus of mandible

Angle of mandible

Coronal suture

Occipitofrontalis muscle
(frontal belly)

Supraorbital margin

Orbicularis oculi muscle

Nasal bone

Zygomatic arch

Masseter muscle

Orbicularis oris muscle

Zygomaticus major muscle

Body of mandible

(b) Right lateral view

FIGURE 2.2 Surface anatomy of the head.

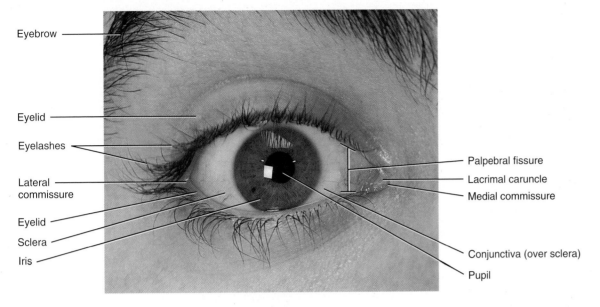

Eyebrow

Eyelid

Eyelashes

Lateral
commissure

Eyelid

Sclera

Iris

Palpebral fissure

Lacrimal caruncle

Medial commissure

Conjunctiva (over sclera)

Pupil

Anterior view

FIGURE 2.3 Surface anatomy of the right eye.

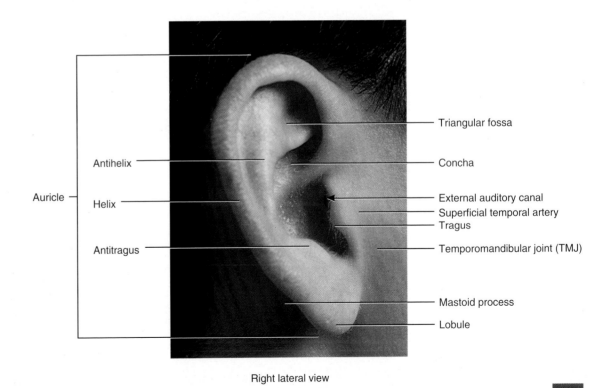

Auricle

Antihelix

Helix

Antitragus

Triangular fossa

Concha

External auditory canal

Superficial temporal artery

Tragus

Temporomandibular joint (TMJ)

Mastoid process

Lobule

Right lateral view

FIGURE 2.4 Surface anatomy of the right ear.

33

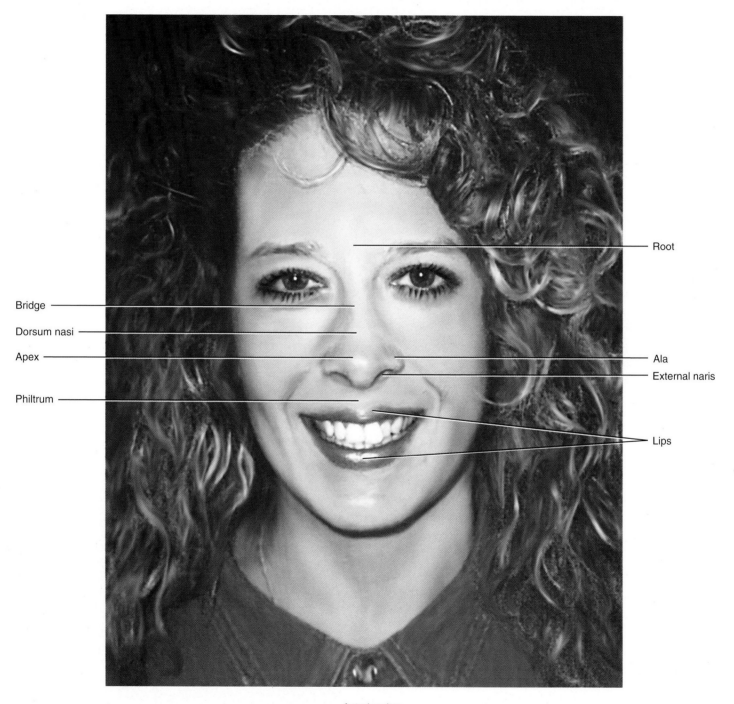

Root

Bridge

Dorsum nasi

Apex

Philtrum

Ala

External naris

Lips

Anterior view

FIGURE 2.5 Surface anatomy of the nose and lips.

34

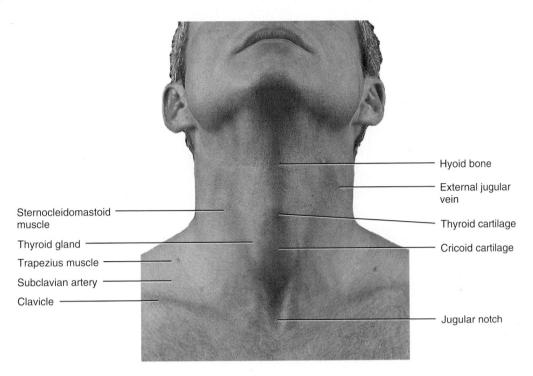

Hyoid bone

External jugular vein

Sternocleidomastoid muscle

Thyroid cartilage

Thyroid gland

Cricoid cartilage

Trapezius muscle

Subclavian artery

Clavicle

Jugular notch

(a) Anterior view

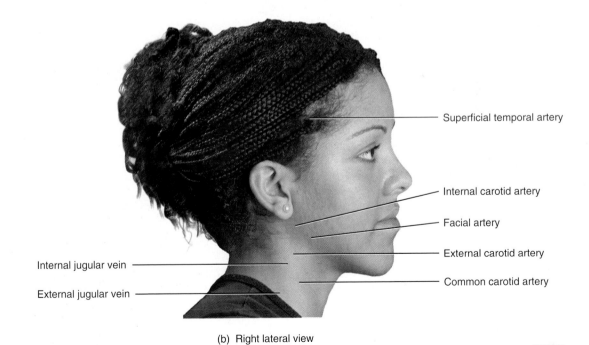

Superficial temporal artery

Internal carotid artery

Facial artery

External carotid artery

Internal jugular vein

Common carotid artery

External jugular vein

(b) Right lateral view

FIGURE 2.6 Surface anatomy of the neck.

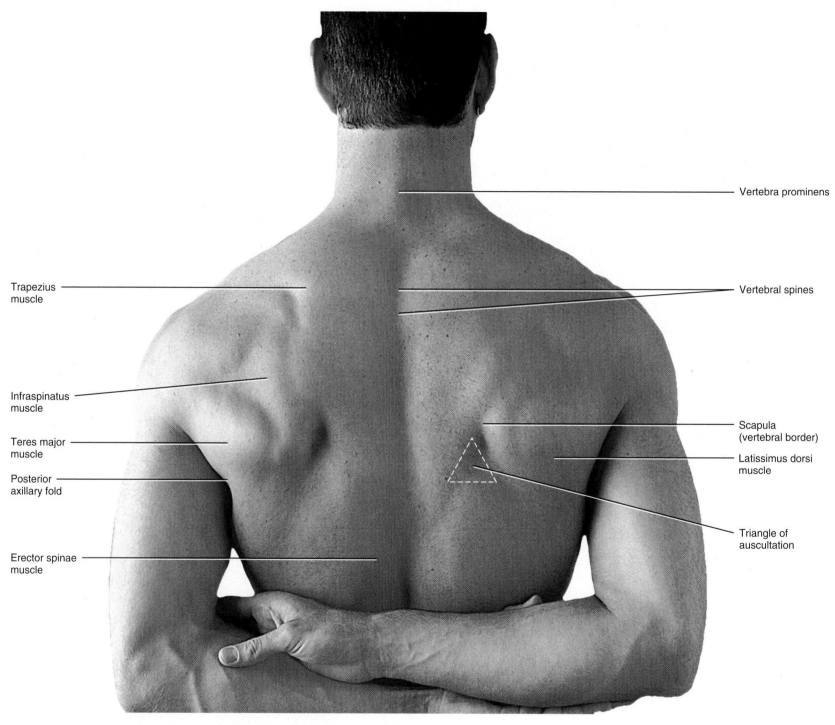

Vertebra prominens

Trapezius
muscle

Vertebral spines

Infraspinatus
muscle

Scapula
(vertebral border)

Teres major
muscle

Latissimus dorsi
muscle

Posterior
axillary fold

Triangle of
auscultation

Erector spinae
muscle

Posterior view

FIGURE 2.7 Surface anatomy of the back.

36

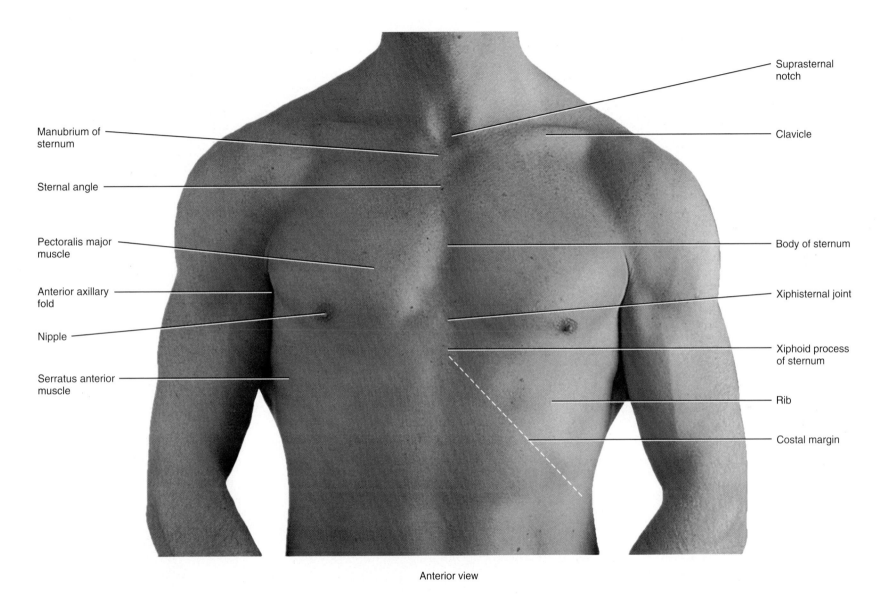

Anterior view

FIGURE 2.8 Surface anatomy of the chest.

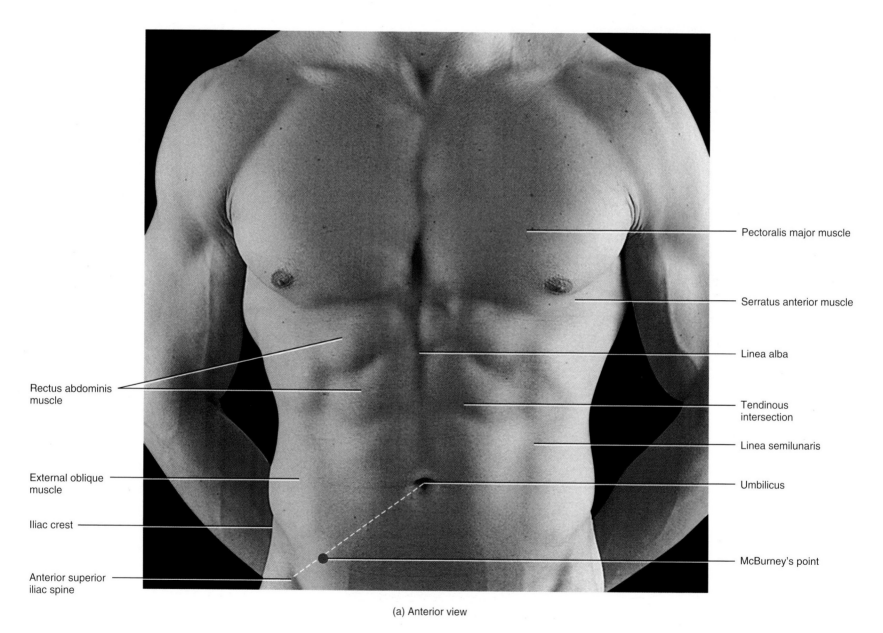

Pectoralis major muscle

Serratus anterior muscle

Linea alba

Rectus abdominis
muscle

Tendinous
intersection

Linea semilunaris

External oblique
muscle

Umbilicus

Iliac crest

Anterior superior
iliac spine

McBurney's point

(a) Anterior view

FIGURE 2.9 Surface anatomy of the abdomen and pelvis.

38

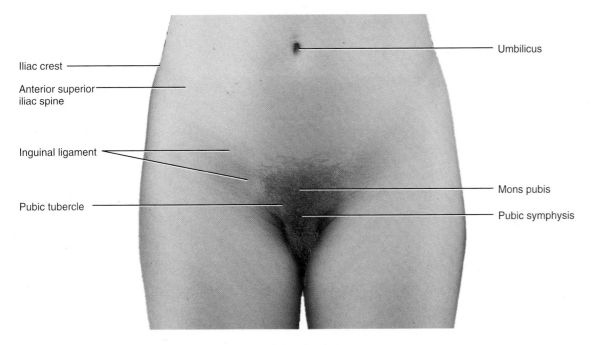

Iliac crest

Anterior superior iliac spine

Inguinal ligament

Pubic tubercle

Umbilicus

Mons pubis

Pubic symphysis

(b) Anterior view

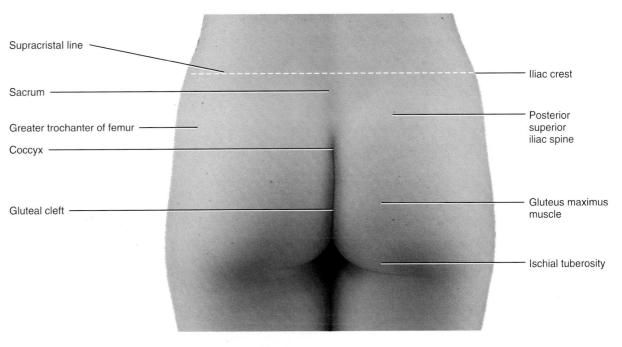

Supracristal line

Sacrum

Greater trochanter of femur

Coccyx

Gluteal cleft

Iliac crest

Posterior superior iliac spine

Gluteus maximus muscle

Ischial tuberosity

(c) Posterior view

FIGURE 2.9 Surface anatomy of the abdomen and pelvis.

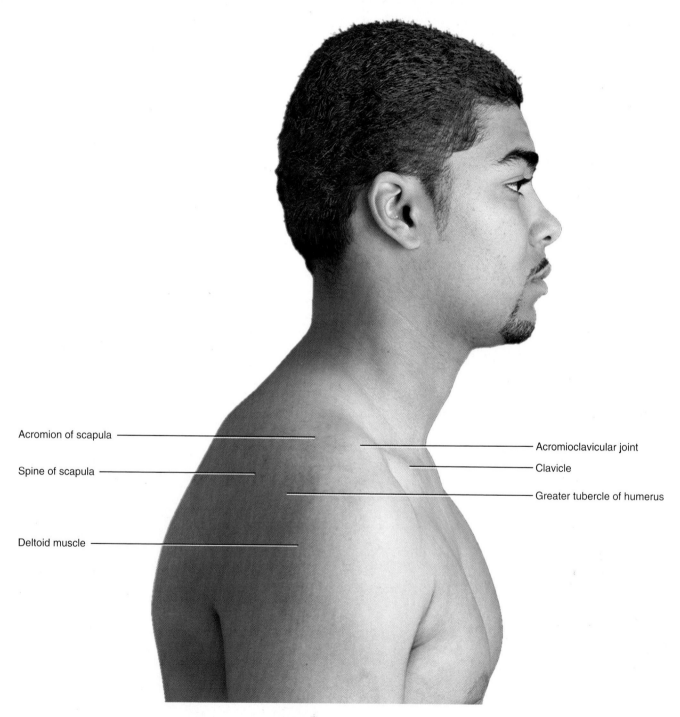

Acromion of scapula ———————————————

————————————— Acromioclavicular joint

Spine of scapula ———————————————

————————————— Clavicle

————————————— Greater tubercle of humerus

Deltoid muscle ———————————————

Right lateral view

FIGURE 2.10 Surface anatomy of the shoulder.

40

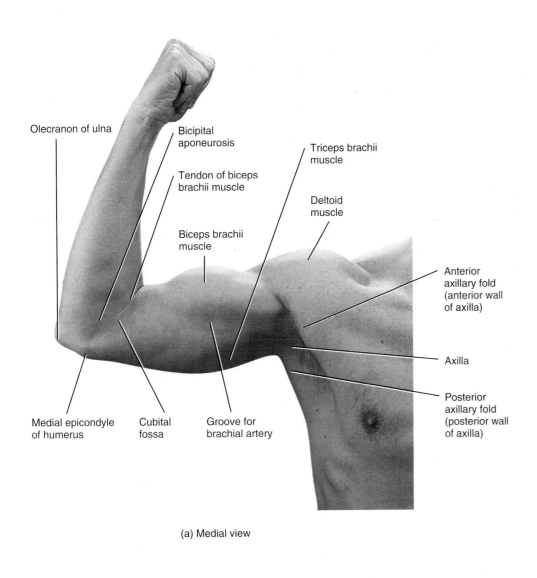

Olecranon of ulna

Bicipital
aponeurosis

Tendon of biceps
brachii muscle

Triceps brachii
muscle

Deltoid
muscle

Biceps brachii
muscle

Anterior
axillary fold
(anterior wall
of axilla)

Axilla

Posterior
axillary fold
(posterior wall
of axilla)

Medial epicondyle
of humerus

Cubital
fossa

Groove for
brachial artery

(a) Medial view

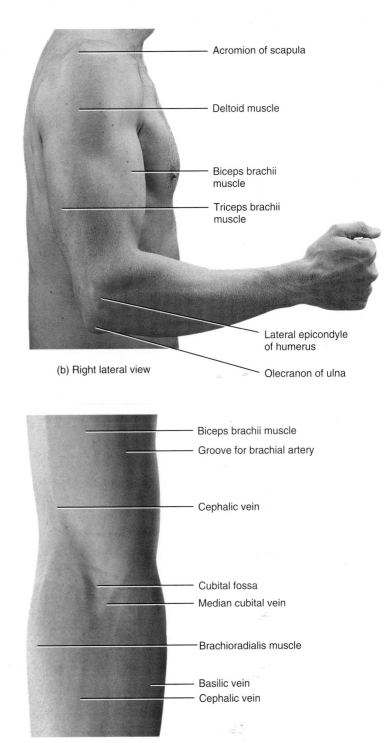

Acromion of scapula

Deltoid muscle

Biceps brachii
muscle

Triceps brachii
muscle

Lateral epicondyle
of humerus

Olecranon of ulna

(b) Right lateral view

Biceps brachii muscle

Groove for brachial artery

Cephalic vein

Cubital fossa

Median cubital vein

Brachioradialis muscle

Basilic vein

Cephalic vein

(c) Anterior view

FIGURE 2.11 Surface anatomy of the arm and elbow.

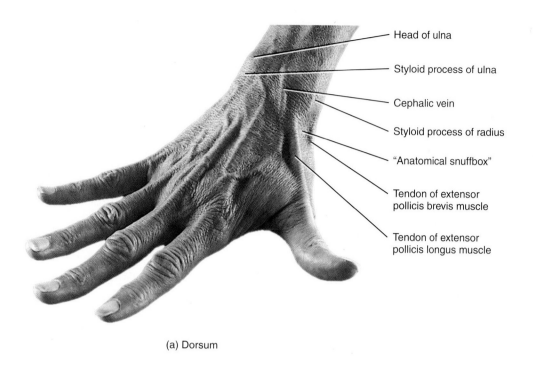

Head of ulna

Styloid process of ulna

Cephalic vein

Styloid process of radius

"Anatomical snuffbox"

Tendon of extensor
pollicis brevis muscle

Tendon of extensor
pollicis longus muscle

(a) Dorsum

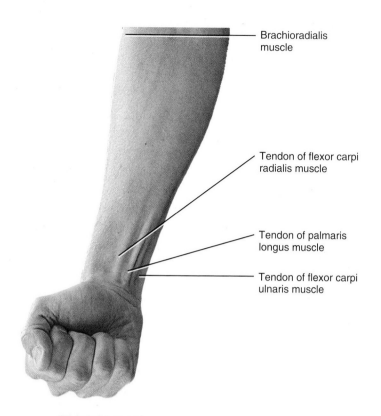

Brachioradialis
muscle

Tendon of flexor carpi
radialis muscle

Tendon of palmaris
longus muscle

Tendon of flexor carpi
ulnaris muscle

(b) Anterior aspect

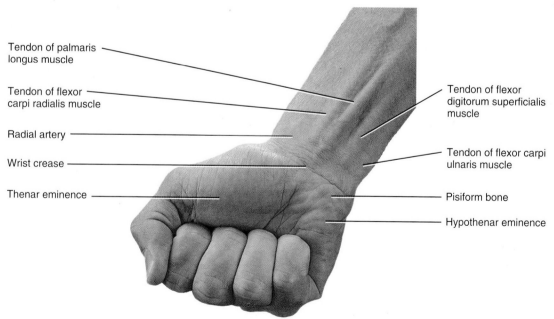

Tendon of palmaris
longus muscle

Tendon of flexor
carpi radialis muscle

Radial artery

Wrist crease

Thenar eminence

Tendon of flexor
digitorum superficialis
muscle

Tendon of flexor carpi
ulnaris muscle

Pisiform bone

Hypothenar eminence

(c) Anterior aspect

FIGURE 2.12 Surface anatomy of the forearm and wrist.

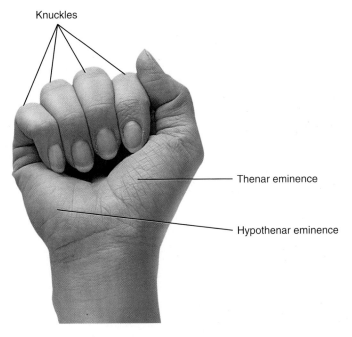

Knuckles

Thenar eminence

Hypothenar eminence

(a) Palmar and dorsal view

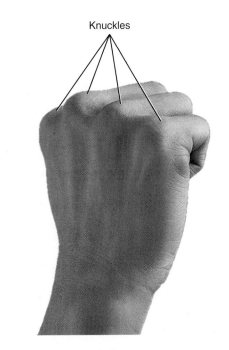

Knuckles

(b) Dorsal view

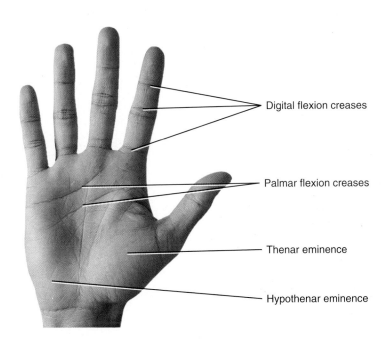

Digital flexion creases

Palmar flexion creases

Thenar eminence

Hypothenar eminence

(c) Palmar view

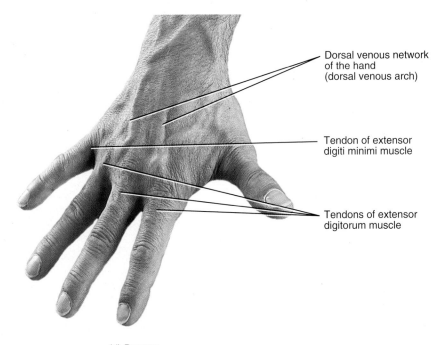

Dorsal venous network
of the hand
(dorsal venous arch)

Tendon of extensor
digiti minimi muscle

Tendons of extensor
digitorum muscle

(d) Dorsum

FIGURE 2.13 Surface anatomy of the hand.

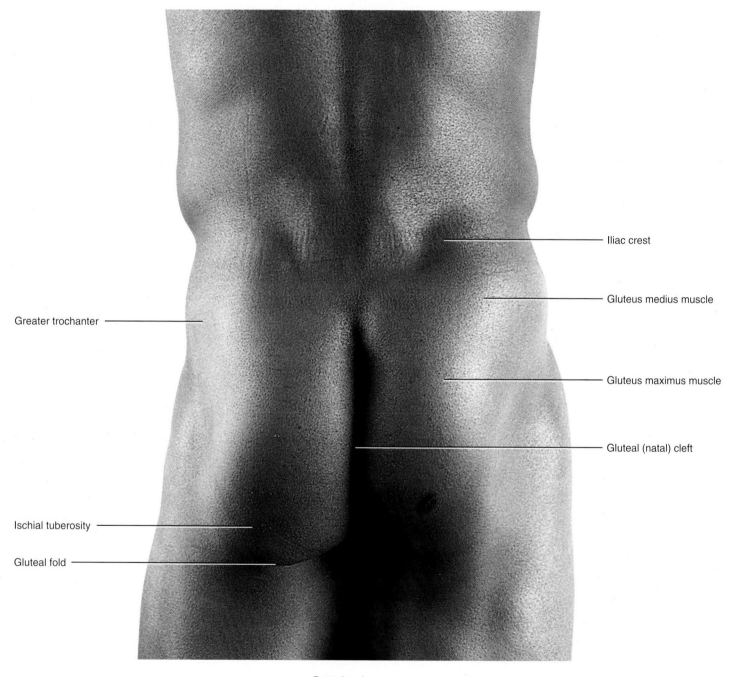

Iliac crest

Gluteus medius muscle

Greater trochanter

Gluteus maximus muscle

Gluteal (natal) cleft

Ischial tuberosity

Gluteal fold

Posterior view

FIGURE 2.14 Surface anatomy of the buttock.

44

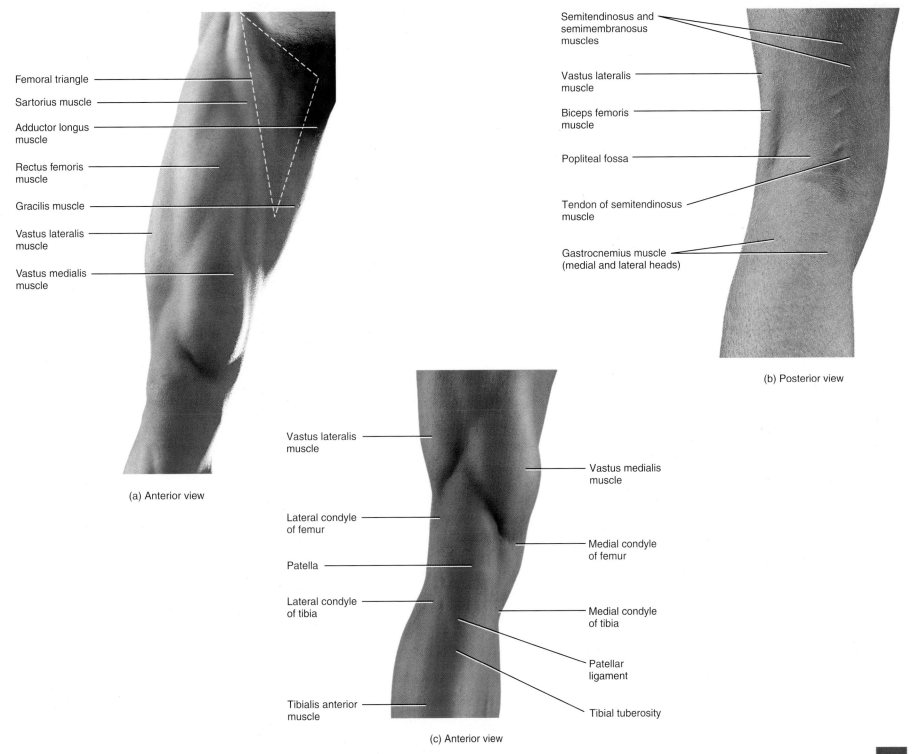

Femoral triangle

Sartorius muscle

Adductor longus
muscle

Rectus femoris
muscle

Gracilis muscle

Vastus lateralis
muscle

Vastus medialis
muscle

(a) Anterior view

Semitendinosus and
semimembranosus
muscles

Vastus lateralis
muscle

Biceps femoris
muscle

Popliteal fossa

Tendon of semitendinosus
muscle

Gastrocnemius muscle
(medial and lateral heads)

(b) Posterior view

Vastus lateralis
muscle

Vastus medialis
muscle

Lateral condyle
of femur

Medial condyle
of femur

Patella

Lateral condyle
of tibia

Medial condyle
of tibia

Patellar
ligament

Tibialis anterior
muscle

Tibial tuberosity

(c) Anterior view

FIGURE 2.15 Surface anatomy of the thigh and knee.

45

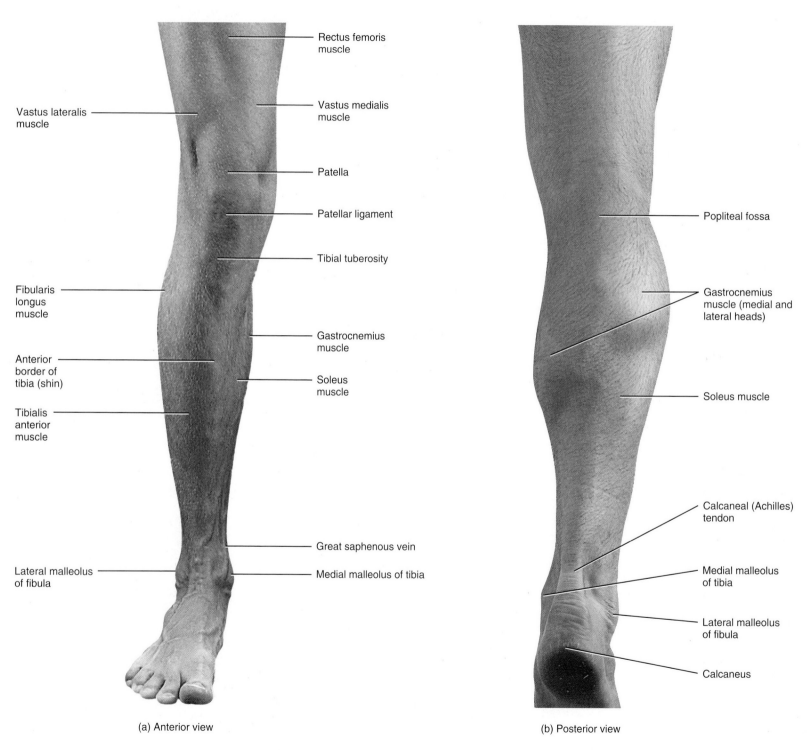

Rectus femoris muscle

Vastus medialis muscle

Patella

Patellar ligament

Tibial tuberosity

Gastrocnemius muscle

Soleus muscle

Great saphenous vein

Medial malleolus of tibia

Vastus lateralis muscle

Fibularis longus muscle

Anterior border of tibia (shin)

Tibialis anterior muscle

Lateral malleolus of fibula

Popliteal fossa

Gastrocnemius muscle (medial and lateral heads)

Soleus muscle

Calcaneal (Achilles) tendon

Medial malleolus of tibia

Lateral malleolus of fibula

Calcaneus

(a) Anterior view

(b) Posterior view

FIGURE 2.16 Surface anatomy of the leg, ankle, and foot.

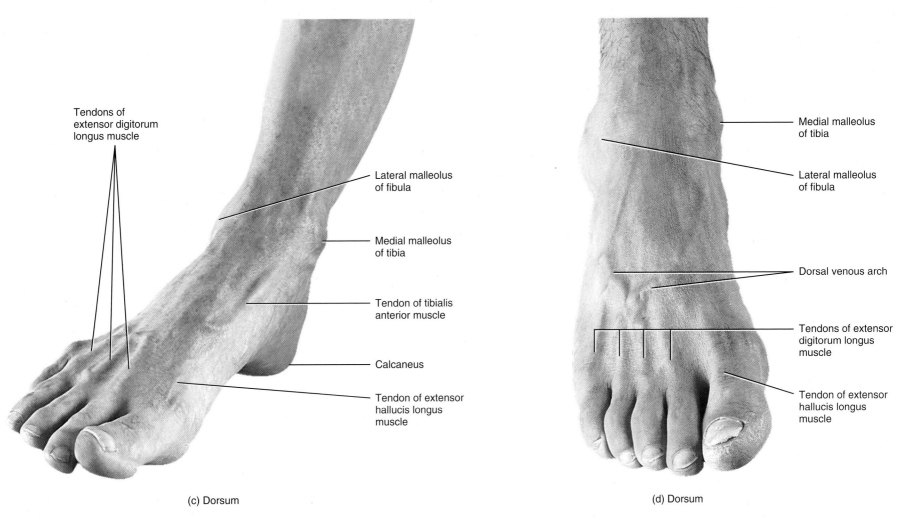

Tendons of
extensor digitorum
longus muscle

Lateral malleolus
of fibula

Medial malleolus
of tibia

Tendon of tibialis
anterior muscle

Calcaneus

Tendon of extensor
hallucis longus
muscle

(c) Dorsum

Medial malleolus
of tibia

Lateral malleolus
of fibula

Dorsal venous arch

Tendons of extensor
digitorum longus
muscle

Tendon of extensor
hallucis longus
muscle

(d) Dorsum

FIGURE 2.16 Surface anatomy of the leg, ankle, and foot.

PART THREE:

MEDICAL IMAGES

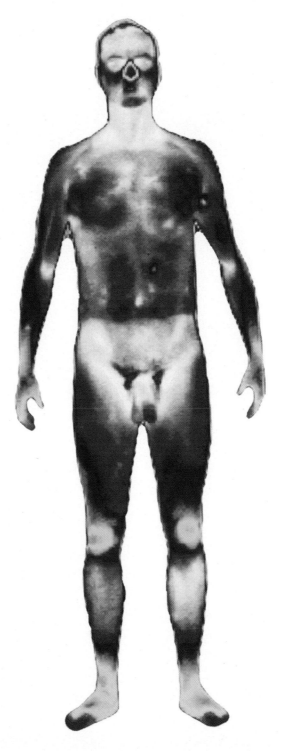

FIGURE 3.1 Thermogram of the male body in anterior view. Different colors represent different temperatures: white (hottest) to red, yellow, green, light blue, dark blue, purple, black (coolest).

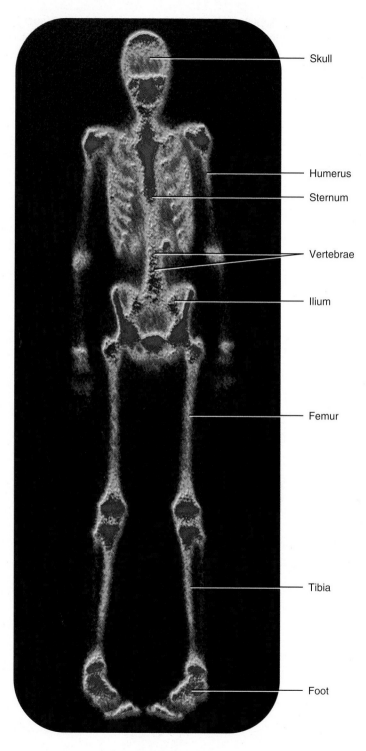

Skull

Humerus

Sternum

Vertebrae

Ilium

Femur

Tibia

Foot

FIGURE 3.2 Radionuclide scan of entire skeleton in anterior view.
Areas of intense color represent high tissue activity.

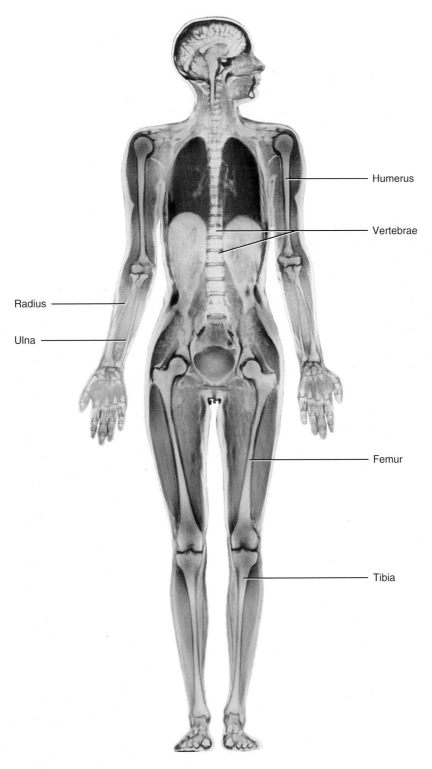

Humerus

Vertebrae

Radius

Ulna

Femur

Tibia

FIGURE 3.3 MRI of female body in frontal section (the head is in sagittal section).

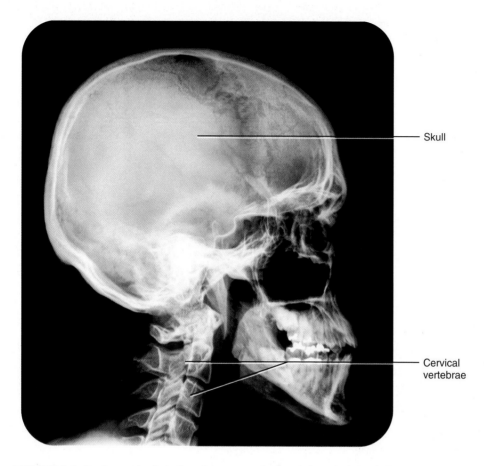

FIGURE 3.4 Radiograph of skull and some cervical vertebrae in lateral view.

Skull

Cervical vertebrae

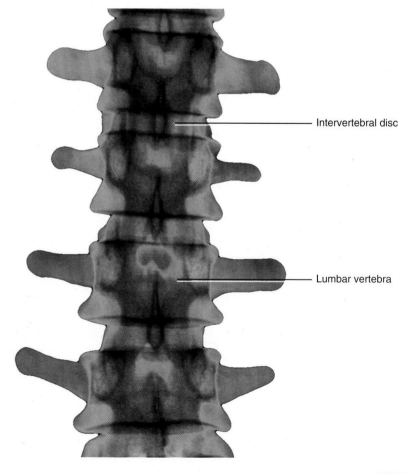

Intervertebral disc

Lumbar vertebra

FIGURE 3.5 Radiograph of lumbar vertebrae in anterior view with their associated intervertebral discs.

54

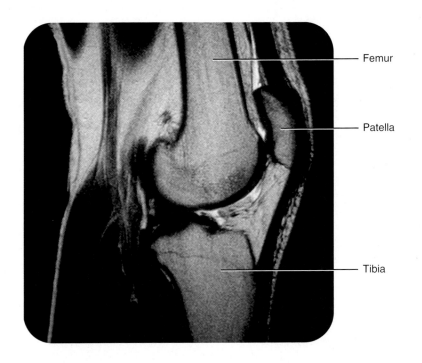

Femur

Patella

Tibia

FIGURE 3.6 MRI of knee joint in sagittal section.

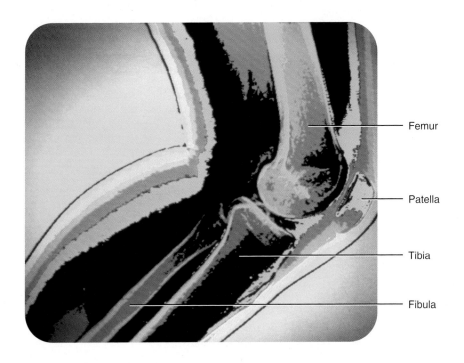

Femur

Patella

Tibia

Fibula

FIGURE 3.7 CT scan of knee in sagittal section.

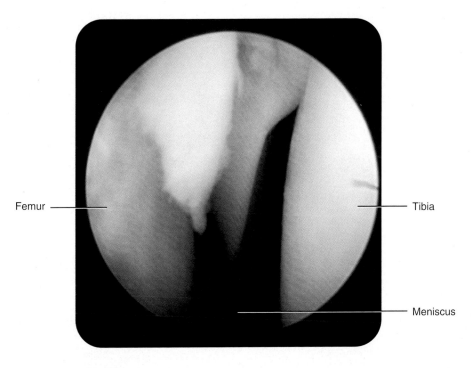

Femur

Tibia

Meniscus

FIGURE 3.8 Endoscopic view of the inside of the knee showing meniscus (articular disc).

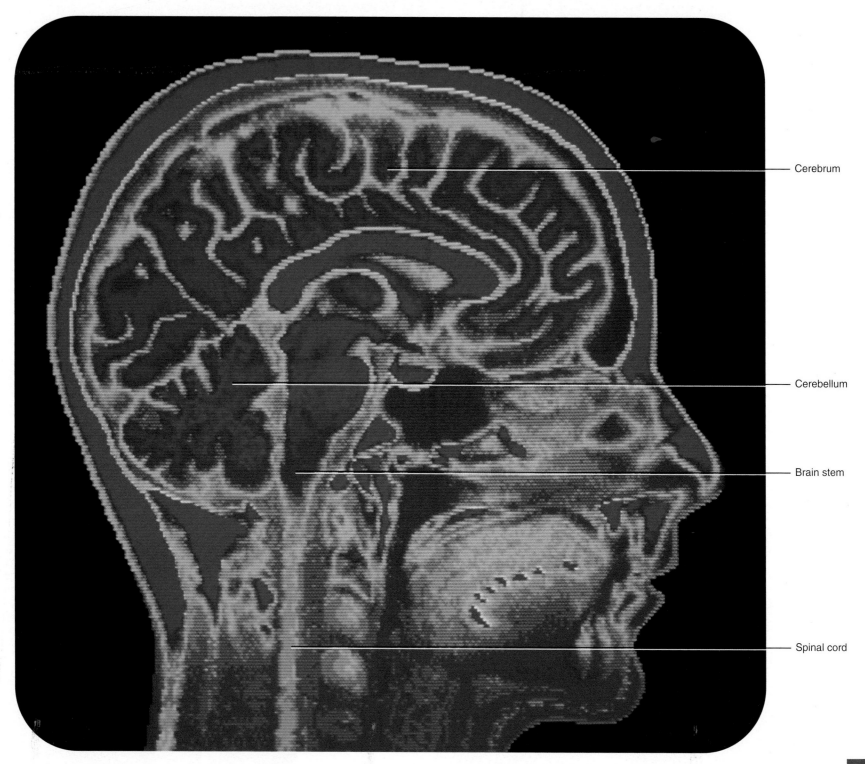

Cerebrum

Cerebellum

Brain stem

Spinal cord

FIGURE 3.9 MRI of the brain and upper spinal cord in midsagittal section.

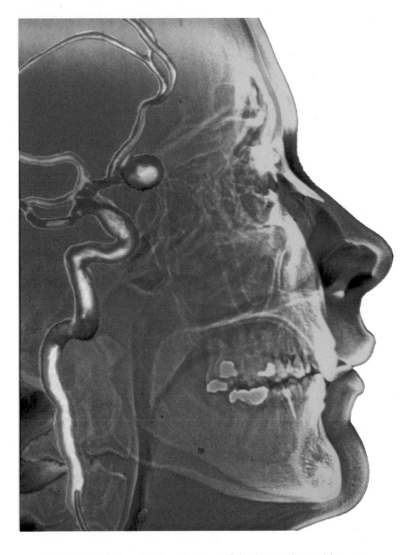

FIGURE 3.10 Cerebral angiogram of the internal carotid artery and some of its major branches. The radiograph of the skull provides orientation.

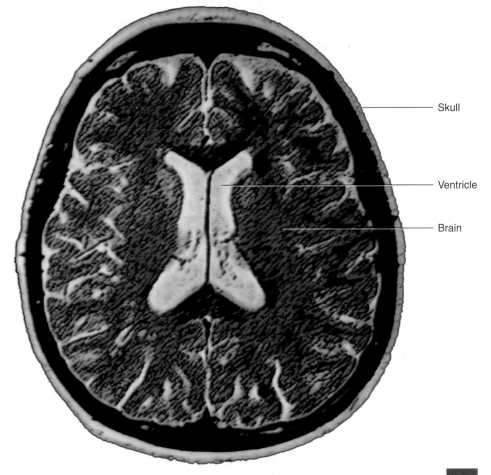

Skull

Ventricle

Brain

FIGURE 3.11 MRI of brain in transverse section.

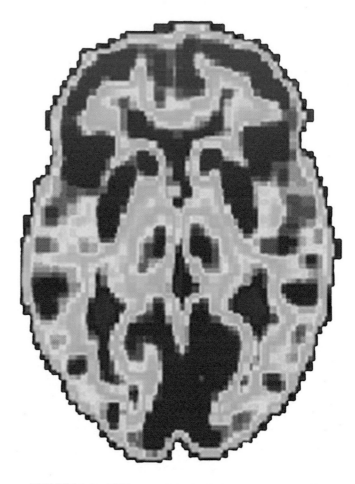

FIGURE 3.12 PET scan of the brain in transverse section.

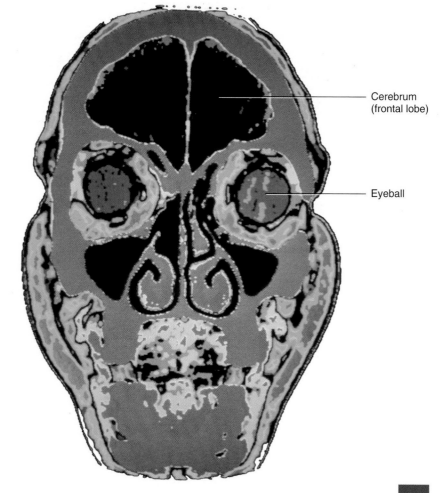

Cerebrum
(frontal lobe)

Eyeball

FIGURE 3.13 CT Scan of brain in frontal section.

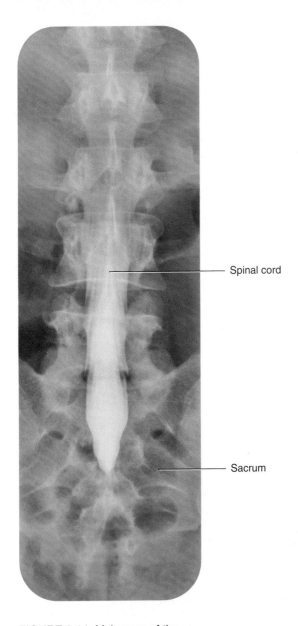

Spinal cord

Sacrum

FIGURE 3.14 Mylogram of the spinal cord in anterior view.

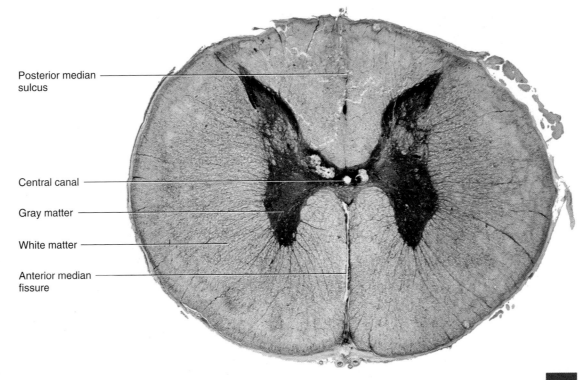

Posterior median sulcus

Central canal

Gray matter

White matter

Anterior median fissure

FIGURE 3.15 Light micrograph of spinal cord in transverse section.

59

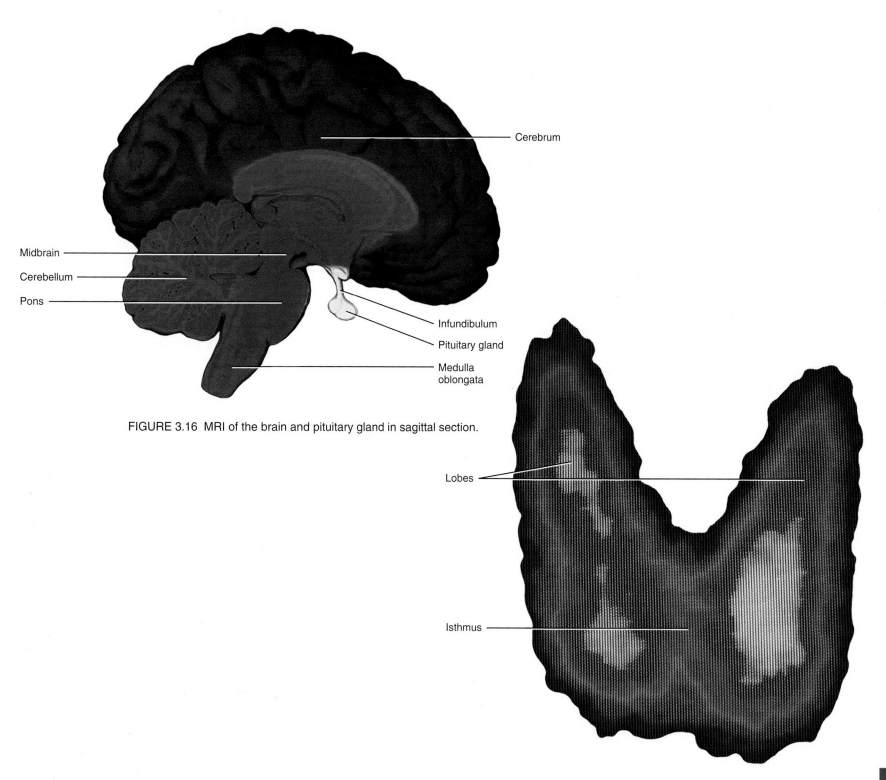

Cerebrum

Midbrain

Cerebellum

Pons

Infundibulum

Pituitary gland

Medulla
oblongata

FIGURE 3.16 MRI of the brain and pituitary gland in sagittal section.

Lobes

Isthmus

FIGURE 3.17 Radionuclide scan of the thyroid gland in anterior view.

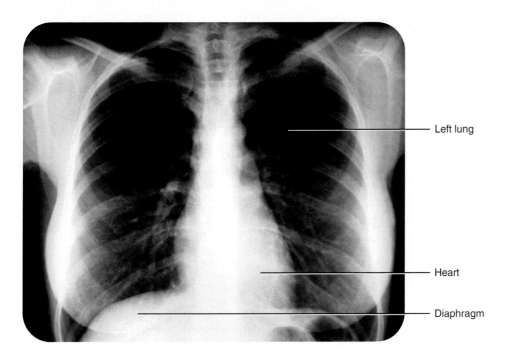

Left lung

Heart

Diaphragm

FIGURE 3.18 Radiograph of the heart in anterior view.

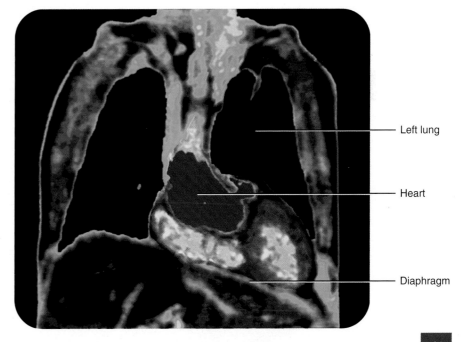

Left lung

Heart

Diaphragm

FIGURE 3.19 MRI scan of the heart in frontal section.

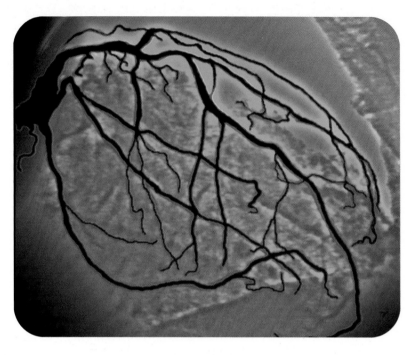

FIGURE 3.20 Angiogram of coronary blood vessels.

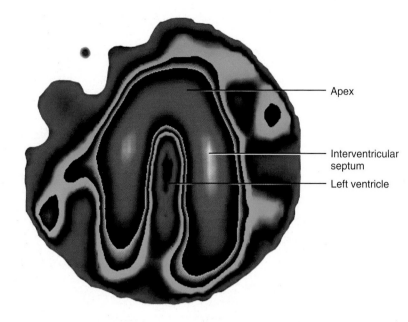

Apex

Interventricular septum

Left ventricle

FIGURE 3.21 SPECT scan of the myocardium in transverse section.

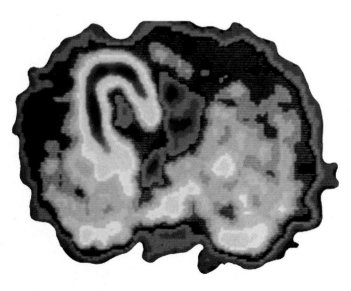

FIGURE 3.22 PET scan of the heart.

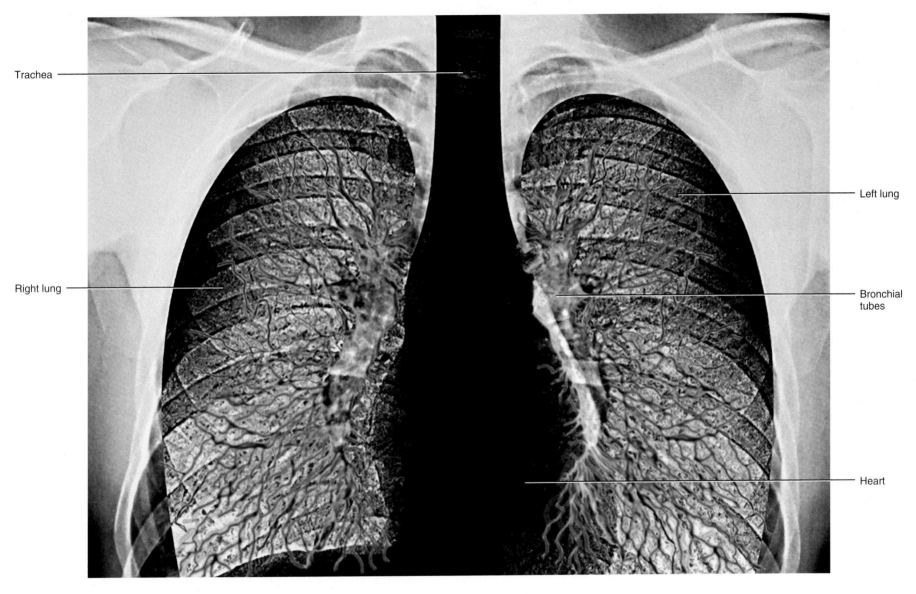

Trachea

Right lung

Left lung

Bronchial tubes

Heart

FIGURE 3.23 X-ray of normal lungs.

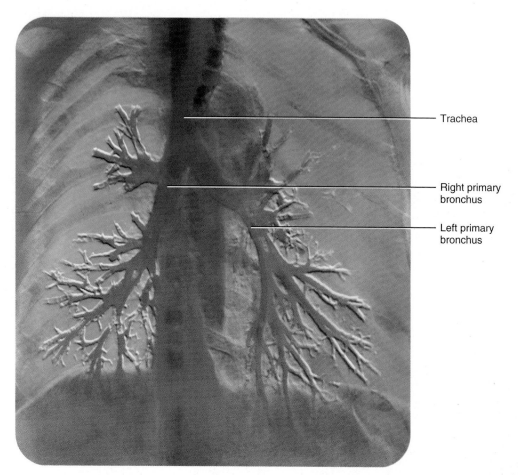

Trachea

Right primary
bronchus

Left primary
bronchus

FIGURE 3.24 Bronchogram of the bronchial tree.

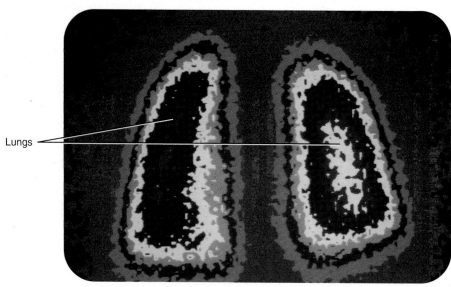

Lungs

FIGURE 3.25 Radionuclide scan of lungs in anterior view.

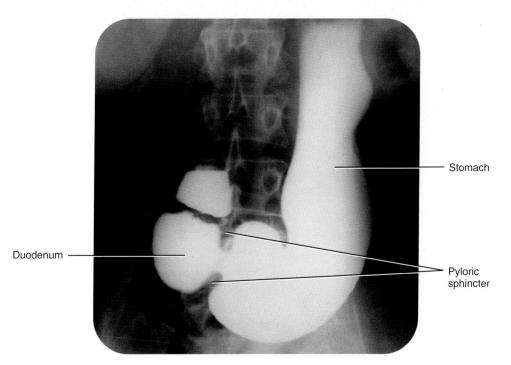

FIGURE 3.26 Radiograph of the stomach and duodenum in anterior view following barium swallow.

Stomach

Duodenum

Pyloric sphincter

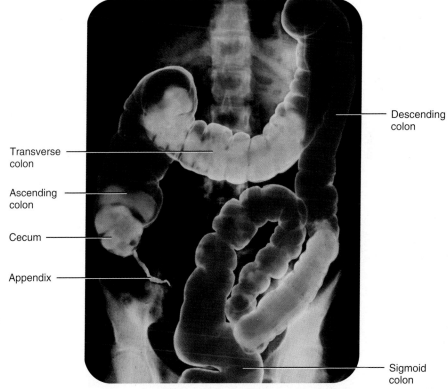

Transverse colon

Ascending colon

Cecum

Appendix

Descending colon

Sigmoid colon

FIGURE 3.27 Radiograph of the large intestine in anterior view.

65

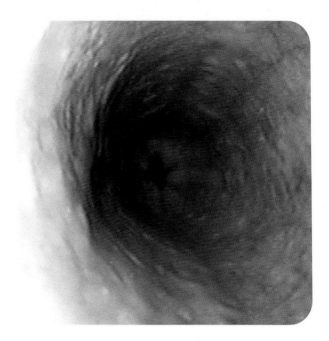

FIGURE 3.28 Endoscopic view of the lining of the esophagus.

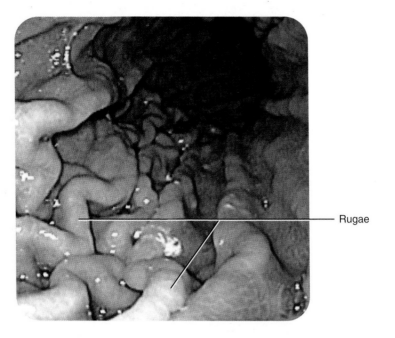

Rugae

FIGURE 3.29 Endoscopic view of the lining of the stomach.

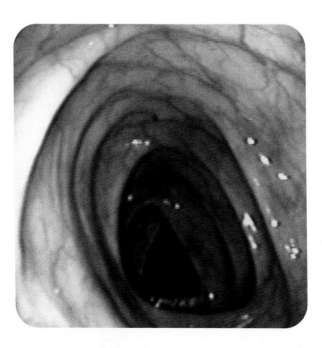

FIGURE 3.30 Endoscopic view of the lining of the colon.

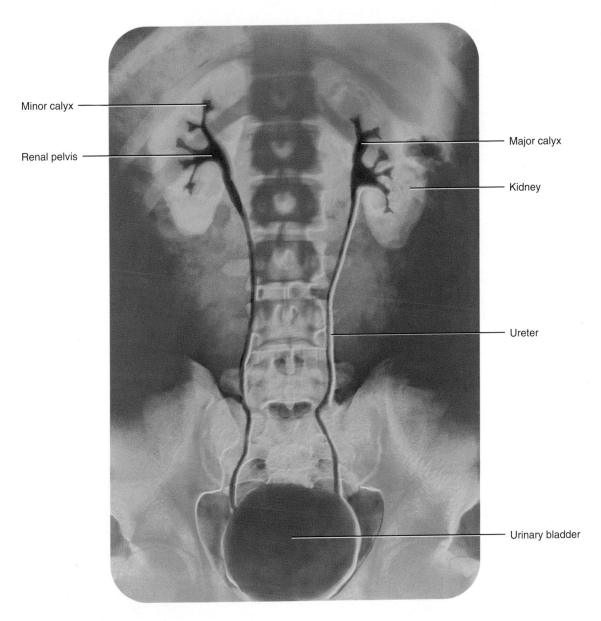

Minor calyx

Renal pelvis

Major calyx

Kidney

Ureter

Urinary bladder

FIGURE 3.31 Radiograph of the urinary system in anterior view.

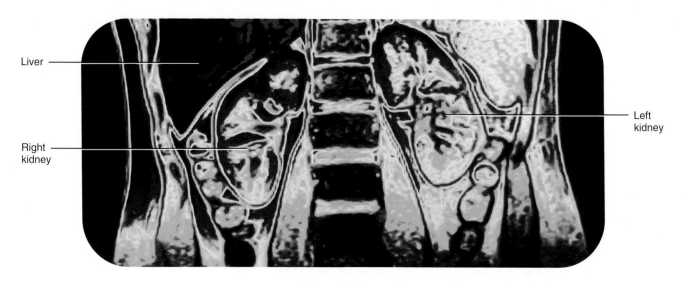

Liver

Right
kidney

Left
kidney

FIGURE 3.32 MRI of the kidneys in frontal section.

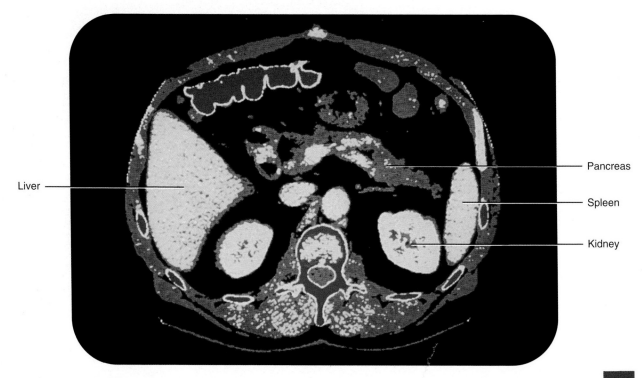

Pancreas

Liver

Spleen

Kidney

FIGURE 3.33 CT scan of the abdomen in transverse section and showing the kidneys in inferior view.

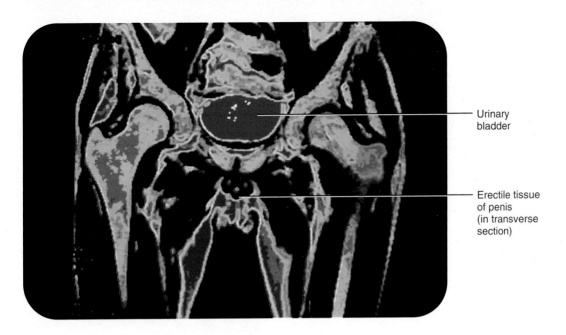

Urinary bladder

Erectile tissue of penis (in transverse section)

FIGURE 3.34 MRI of male pelvis in frontal section.

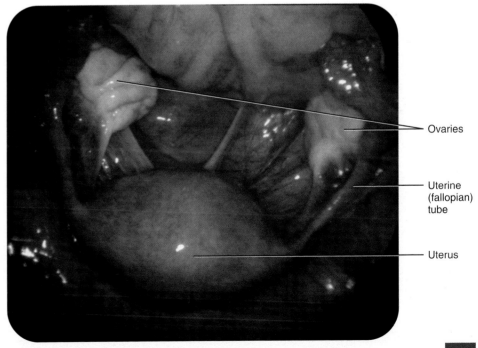

Ovaries

Uterine (fallopian) tube

Uterus

FIGURE 3.35 Laparoscopy of female abdominopelvic cavity.

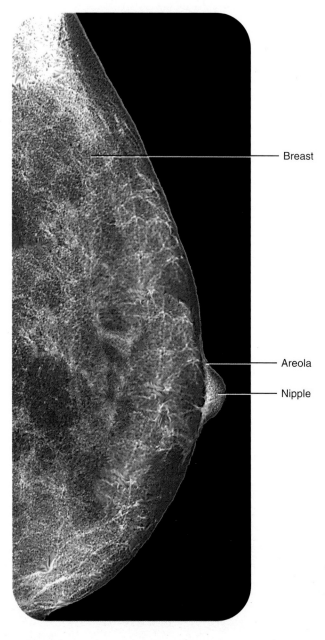

Breast

Areola

Nipple

FIGURE 3.36 Mammogram in lateral view.

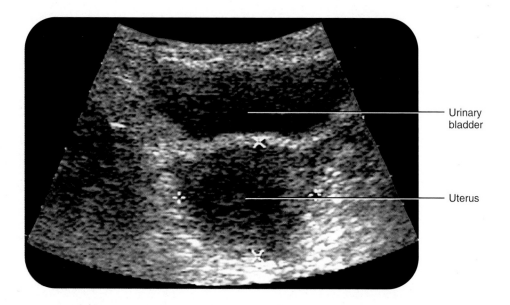

Urinary
bladder

Uterus

FIGURE 3.37 Sonogram of the uterus and urinary bladder.

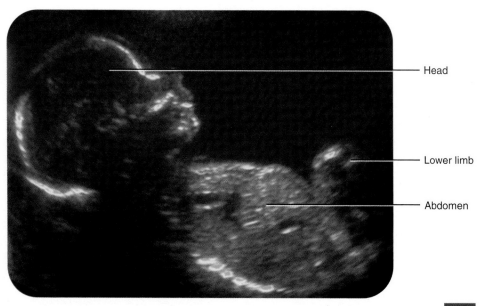

Head

Lower limb

Abdomen

FIGURE 3.38 Ultrasound image of six-month fetus.

PHOTO CREDITS